THE REIGN OF LAW.

Let knowledge grow from more to more,
But more of reverence in us dwell;
That mind and soul, according well,
May make one music as before,

But vaster . . . . .

In Memoriam.

THE REIGN OF LAW.

By the DUKE OF ARGYLL.

FOURTH AMERICAN EDITION.

NEW YORK:
GEORGE ROUTLEDGE & SONS,
416 BROOME STREET.
1873.

PREFACE TO THE FIFTH EDITION.

IN preparing a Fifth Edition of this work, I have to acknowledge the favour — far greater than I expected—with which it has been received. The argument which it maintains is at variance with the philosophy of some of the most active and popular thinkers of the time; and on a few important points it deviates from the view commonly adopted by men with whom I am more generally agreed. Some adverse comment was therefore not only to be expected but desired. Most sincerely do I thank those who, in numerous Journals and Reviews, have undertaken this duty, for the uniformly courteous and even kindly spirit in which their criticisms have been expressed.

In this Edition no alteration has been made involving any change of principle or opinion. Here and there words have been added or removed according as individual passages appear to have been misunderstood. Throughout some of the chapters substantial additions have been made in reply, direct or indirect, to my principal opponents, whilst discussions, more detailed than were suitable for the text, have been committed to Notes at the end of the volume.

These additions and Notes have reference chiefly to the following articles which appeared in review of the "Reign of Law:"—

1st. An Article, by Mr. Alfred R. Wallace, in the *Quarterly Journal of Science*, for October 1867. This article is in defence and illustration of Mr. Darwin's "Theory on the Origin of Species." The eminence of Mr. Wallace as a Naturalist, the extent of his researches in some of the most remarkable Faunas of the world, and the fact that, before the publication of Mr. Darwin's book, he had come to kindred, if not identical conclusions,—all render him peculiarly competent to defend the "Theory," and to present it in the strongest light. I have therefore added to the text several passages suggested by the challenge he makes, and by the reasoning he employs. A further discussion of his paper will be found in Note A.

2d. An Article, by Mr. George H. Lewes, in the *Fortnightly Review*, for July, 1867, dealing with the main argument and conclusion of this work from the well-known point of view of the "Positive Philosophy." Wherever in the text there seemed a fitting place for doing so, I have inserted passages which deal with the reasoning of his paper, or with the same reasoning as it appears in a more systematic form in the Prolegomena to Mr. Lewes's "History of Philosophy."

3d. An Article in the *Dublin Review*, for April 1867, which I am permitted to attribute to the learned editor of that periodical, Dr. Ward. The more special object of his adverse comment is the view I have taken of the doctrine of Free Will—a doctrine which Dr. Ward, with some warmth, accuses me of having virtually abandoned whilst professing to defend it. A slight alteration in the text may perhaps help to remove some objections, which rest entirely upon a misunderstanding of the sense in which particular words are used. But behind and beyond any misunderstanding of this kind, there lies apparently a substantial difference in respect to which my view remains unaltered. This difference will be found discussed in Note F, at the end of the volume.

4th. An Article in the *Contemporary Review*, for May 1867, by Mr. J. P. Mahaffy. With reference to his observations, as well as to those of some other critics, I have somewhat expanded several passages which deal with the Supernatural, and with the various relations in which miracles have been conceived to stand towards the "Reign of Law." I have also, in a special Note (G), replied to a criticism in this paper, referring to the subject of Necessity and Free Will.

Other Notes have been added in illustration or support of various passages in the text.

As regards the intention I had at one time entertained of adding a chapter on "Law in Christian Theology," further reflection has only confirmed me in the feeling that this is a subject which cannot be adequately dealt with in such a form. I can only again ask my readers to remember that although some ideas which belong to this subject, or touch it at various points, cannot be, and have not been, avoided, yet the desire and intention to postpone it, in so far as it was possible to do so, has left blanks which every careful eye must see.

INVERARAY, *Jan.* 1868.

PREFACE TO THE FIRST EDITION.

SOME portions of this work have already appeared at various times in the *Edinburgh Review*, in *Good Words*, and in Addresses to the Royal Society of Edinburgh during the years in which I had the honour of being President of that Body. The deep interest of the matter dealt with in those Papers has induced me to expand them, to add new chapters on other aspects of the same subject, and to publish the whole in a connected form.

Among many other deficiencies which may be observed in this Volume, there is one which demands explanation, lest a serious misunderstanding should arise. I had intended to conclude with a chapter on "Law in Christian Theology." It was natural to reserve for that chapter all direct reference to some of the most fundamental facts of Human nature. Yet without such reference the Reign of Law, especially in the "Realm of Mind," cannot even be approached in some of its very highest and most important aspects. For the present, however, I have shrunk from entering upon questions so profound, of such critical import, and so inseparably connected with religious controversy. In the absence of any attempt to deal with this great branch of the inquiry, as well as in many other ways, I am painfully conscious of the narrow range of this work. I can only offer it as a very small contribution to the discussion of a boundless subject.

INVERARAY *October*, 1866.

CONTENTS.

CHAPTER I.

THE SUPERNATURAL.

CHAPTER II.

LAW: ITS DEFINITIONS.

CHAPTER III.

CONTRIVANCE A NECESSITY ARISING OUT OF THE REIGN OF LAW.

CHAPTER IV.

APPARENT EXCEPTIONS TO THE SUPREMACY OF PURPOSE.

CHAPTER V.

CREATION BY LAW.

CHAPTER VI.

THE REIGN OF LAW IN THE REALM OF MIND.

CHAPTER VII.

LAW IN POLITICS.

LIST OF ILLUSTRATIONS.

THE REIGN OF LAW.

CHAPTER I.

THE SUPERNATURAL.

THE Supernatural—what is it? What do we mean by it? How do we define it? M. Guizot[1] tells us that belief in it is the special difficulty of our time—that denial of it is the form taken by all modern assaults on Christian faith; and again, that acceptance of it lies at the root, not only of Christianity, but of all positive religion whatever. These questions, then, concerning the Supernatural, are questions of first importance. Yet we find them seldom distinctly put, and still more seldom distinctly answered. This is a capital error in dealing with any question of philosophy. Half the perplexities of men are traceable to obscurity of thought hiding and breeding under obscurity of language. "The Super-

[1] "L'Église et la Société Chretienne en 1861," ch. iv. p. 19.

natural" is a term employed often in different, and sometimes in contradictory, senses. It is difficult to make out whether M. Guizot himself means to identify belief in the Supernatural with belief in the existence of a God, or with belief in a particular mode of Divine action. But these are ideas quite separable and distinct. There may be some men who disbelieve in the Supernatural only because they are absolute atheists; but it is certain that there are others who have great difficulty in believing in the Supernatural, who are not atheists. What they doubt or deny is, not that God exists, but that He ever acts, or perhaps can act, unless in and through what they call the "Laws of Nature." M. Guizot, indeed, tells us that "God is the Supernatural in a Person." But this is a rhetorical figure rather than a definition. He may, indeed, contend that it is inconsistent to believe in a God, and yet to disbelieve in the Supernatural; but he must admit, and indeed does admit, that such inconsistency is found in fact.

Theological and philosophical writers frequently use the Supernatural as synonymous with the Superhuman. But of course this is not the sense in which any one can have any difficulty in believing in it. The powers and works of Nature are all superhuman—more than Man can account for in their origin—more than he can resist in their energy—more than he can understand in their effects. This, then, cannot be the sense in which so many minds find it hard to accept the Supernatural; nor can it be the sense in which others cling to it as of the very essence of their religious faith. What, then, is that other sense in which the difficulty arises? Perhaps we shall best find it by seeking the idea which is competing with it, and by which it has been displaced. It is the Natural which has been casting out the Supernatural—the idea of Natural Law,—the universal reign of a fixed Order of things. This idea is a product of that immense development of the physical sciences which is characteristic of our time. We cannot read a periodical, or go into a lecture-room, without hearing it expressed. Sometimes, but rarely, it is stated with accuracy, and with due recognition of the limits within which Law can be said to comprehend the phenomena of the world. But generally it is expressed in language vague and hollow, covering inaccurate conceptions, and confounding under common forms of expression ideas which are essentially distinct. The mere ticketing and orderly assortment of external facts is constantly spoken of as if it were in the nature of Explanation, and as if no higher truth in respect to natural phenomena were to be attained or desired.[1] And herein we see both the result for which

[1] Those who have followed the course of recent speculation will recognise this sentence as intended to describe the characteristic

Bacon laboured, and the danger against which Bacon prayed. It has been a glorious result of a right method in the study of Nature, that with the increase of knowledge the "human family has been endowed with new mercies." But every now and then, for a time at least, from "the unlocking of the gates of sense, and the kindling of a greater natural light, incredulity and intellectual night *have* arisen in our minds."[1]

But let us observe exactly where and how the difficulty arises. The Reign of Law in Nature is, indeed, so far as we can observe it, universal. But the common idea of the Supernatural is that which is at variance with Natural Law, above it, or in violation of it. Nothing, however wonderful, which happens according to Natural Law, would be considered by any one as Supernatural. The law in obedience to which a wonderful thing happens may not be known; but this would not give it a supernatural character, so long as we assuredly believe

principle of the Positive Philosophy. I am glad to observe that so competent a judge as Mr. George H. Lewes says of it:—"Although not, perhaps, the most dignified or explicit statement of the Positive point of view, this may be accepted as essentially correct."—*Fortnightly Review*, July 1867.

1 "This also we humbly beg, that human things may not prejudice such as are Divine, neither that from the unlocking of the gates of sense, and the kindling of a greater natural light, anything of incredulity or intellectual night may arise in our minds towards Divine mysteries."—"The Student's Prayer." *Bacon's Works.*

that it did happen according to *some* law. Hence, it would appear to follow that a man thoroughly possessed of the idea of Natural Law as universal, never could admit anything to be supernatural; because on seeing any fact, however new, marvellous, or incomprehensible, he would escape into the conclusion that it was the result of some natural Law of which he had before been ignorant. No one will deny that, in respect to the vast majority of all new and marvellous phenomena, this would be the true and reasonable conclusion. It is not the conclusion of pride, but of humility of mind. Seeing the boundless extent of our ignorance of the natural laws which regulate so many of the phenomena around us, and still more of so many of the phenomena within us, nothing can be more reasonable than to conclude, when we see something which is to us a wonder, that somehow, if we only knew how, it is "all right"—all according to the constitution and course of Nature. But then, to justify this conclusion, we must understand Nature in the largest sense,—as including all that is

> "In the round ocean, and the living air
> And the blue sky, *and in the mind of man*."[1]

We must understand it as including every agency which we see entering, or can conceive from analogy as capable

[1] "Tintern Abbey."—Wordsworth.

of entering, into the causation of the world. First and foremost among these is the agency of our own Mind and Will. Yet, strange to say, all reference to this agency is often tacitly excluded when we speak of the laws of Nature. One of our most distinguished living teachers of physical science[1] began, not long ago, a course of lectures on the phenomena of Heat by a rapid statement of the modern doctrine of the Correlation of Forces—how the one was convertible into the other—how one arose out of the other—how none could be evolved except from some other as a pre-existing source. "Thus," said the lecturer, "we see there is no such thing as spontaneousness in Nature." What !—not in the lecturer himself? Was there no "spontaneousness" in his choice of words—in his selection of materials—in his orderly arrangement of experiments with a view to the exhibition of particular results? It is not probable that the lecturer was intending to deny this; it simply was that he did not think of it as within his field of view. His own Mind and Will were then dealing with the "laws of Nature," but they did not occur to him as forming part of those laws, or, in the same sense, as subject to them.

Does Man, then, not belong to Nature? Is he above

[1] Professor Tyndall.

it—or merely separate from it, or a violation of it? Is he supernatural? If so, has he any difficulty in believing in himself? Of course not. Self-consciousness is the one truth, in the light of which all other truths are known. *Cogito, ergo sum*, or *volo, ergo sum*—this is the one conclusion which we cannot doubt, unless Reason disbelieves herself. Why, then, are the faculties of the human mind and body not habitually included among the "laws of Nature?" Because a fallacy is getting hold upon us from a want of definition in the use of terms. "Nature" is being used in the narrow sense of physical nature. It is conceived as containing nothing beyond the properties of Matter. Thus the whole mental world in which we ourselves live, and move, and have our being, is excluded from it. But these selves of ours do belong to Nature. At all events if we are ever to understand the difficulties in the way of believing in the Supernatural, we must first keep clearly in view what we intend to understand as included in the Natural. Let us never forget, then, that the agency of Man is of all others the most natural—the one with which we are most familiar—the only one, in fact, which we can be said, even in any measure, to understand. When any wonderful event can be referred to the contrivance or ingenuity of Man, it is thereby at once removed from the sphere of the Supernatural, as ordinarily understood.

It must be remembered, however, that we are now only seeking a clear definition of terms; and that provided this other meaning be clearly agreed upon, the Mind and Will of Man may be considered as separate from "nature," and belonging to the Supernatural. This view is taken in an able treatise on "Nature and the Supernatural," by Dr. Bushnell, an American clergyman.[1] Dr. Bushnell says:—"That is supernatural, whatever it be, that is either not in the chain of natural cause and effect, or which acts on the chain of cause and effect in nature, from without the chain." Again:—"If the processes, combinations, and results of our system of nature are interrupted or varied by the action, whether of God, or angels, *or men*, so as to bring to pass what would not come to pass in it by its own internal action, under the laws of mere cause and effect, such variations are in like manner supernatural." There is no other objection to this definition of the Supernatural, than that it rests upon a limitation of the terms "Nature" and "natural," which is very much at variance with the sense in which they are commonly understood. There is, indeed, a distinction which finds its expression in common language between the works of Man and the works of Nature. A honeycomb, for example, would be called a work of

[1] "Nature and the Supernatural, as together constituting the one System of God." By Horace Bushnell, D.D. Edinburgh, 1860.

Nature, but a steam-engine would not. This distinction is founded on a true perception of the fact that the Mind and Will of Man belong to an order of existence very different from physical laws, and very different also from the fixed and narrow instincts of the lower animals. It is a distinction bearing witness to the universal consciousness that the Mind of Man has within it something of a truly creative energy and force—that we are in a sense "fellow-workers with God," and have been in a measure "made partakers of the Divine nature." Nevertheless, it would be using the word in a sense very different from that in which it is generally accepted, were we to call the steam-engine a supernatural work. Yet it does answer strictly to the definition of Dr. Bushnell in being "the result of natural Law varied by the action of men." It is made by "acting on the chain of cause and effect in nature from without the chain." But then, be it observed, that under the same definition all the contrivances of Nature become Supernatural the moment they are conceived as the work of a Mind using what we call the elements of nature for the accomplishment of its designs. If, for example, it is open to us to conceive that such a creature as a Bee cannot have been made out of those elements "by their own internal action," then we must regard both this creature and the wonderful products of its instinct as belonging to the Supernatural. The honeycomb and the steam-engine would thus come under the same category—with this only difference, that the mind which made the steam-engine, being connected with a Body, is visibly known to us, whereas the Mind which made the Bee is withdrawn from sight. But both can be equally regarded as the result of Mind "acting on the chain of cause and effect from without the chain." Nor can we stop here. The same process of analysis will carry us farther in the same direction. We often speak, as Dr. Bushnell does here, of the elementary forces of Nature as "acting" by themselves. But there is no other meaning in these words than an expression of the fact that we neither see nor understand the connexion of those elementary forces and Mind. But this ignorance of ours affords no manner of presumption that such connexion does not exist. On the contrary, though the manner of that connexion be unknown, it is much more conceivable to us that some connexion does exist than that it does not. If therefore the distinction between the Natural and the Supernatural be the distinction between that which is and that which is not the work of Mind, then it becomes a purely arbitrary distinction. It assumes that we can distinguish between cases in which the properties of matter work under the direction of Mind, and other cases in which they work "of themselves." But this is a line which we draw for ourselves. There is no reason to suppose that it has any reality in the constitution of things. It is not in those things, but in the point of view from which we regard them, that the distinction lies. We have only to change that point of view, and the distinction vanishes. All Nature becomes Supernatural, because all her elements, both in themselves and in their combinations, are only conceivable as first established, and then employed by the powers of Mind.

But if this definition of the Supernatural displeases us, as tending to confound distinctions which we had thought were clear, let us take another definition. Let us take the Natural in that larger and wider sense, in which it contains within it the whole phenomena of Man's intellectual and spiritual nature, as part, and the most familiar of all parts, of the visible system of things. This is a definition more consonant with common language. In all ordinary senses of the term, Man and his doings belong to the Natural, as distinguished from the Supernatural.

We are now from another point of view coming nearer to some precise understanding of what the Supernatural may be supposed to mean. But before we proceed, there is another question which must be answered—What is the relation in which the agency of Man stands to the physical laws of Nature? The answer, in part at least, is plain. His power in respect to those laws extends only, first to their discovery and ascertainment, and then to their use. He can establish none: he can suspend none. All he can do is to guide, in a limited degree, the mutual action and reaction of the laws amongst each other. They are the tools with which he works—they are the instruments of his Will. In all he does or can do he must employ them. His ability to use them is limited both by his want of knowledge and by his want of power. The more he knows of them, the more largely he can employ them, and make them ministers of his purposes. This, as a general rule, is true; but it is subject to the second limitation just pointed out. Our power over Nature does not necessarily keep pace with our knowledge of her Laws. Man already knows far more than he has power to convert to use. It is a true observation of Sir George Lewis,[1] that Astronomy, for example, in its higher branches, has an interest almost purely scientific. It reveals to our knowledge perhaps the grandest and most sublime of the physical laws of Nature. But a much smaller amount of knowledge would suffice for the only practical applications which we have yet been able to make of these laws to our own use. Still, that knowledge has a reflex

[1] "Astronomy of the Ancients," p. 254.

influence on our knowledge of ourselves, of our powers, and of the relations which subsist between the constitution of our own minds and the constitution of the universe. And in other spheres of inquiry, advancing knowledge of physical laws has been constantly accompanied with advancing power over the physical world. It has enabled us to do a thousand things, any one of which, a few generations ago, would have been considered supernatural. Nor can it be said that this judgment of their character would have been erroneous. These things would have been superhuman then, though they are not superhuman now. The same lecturer who told his audience that there was nothing spontaneous in Nature proceeded, by virtue of his own knowledge of natural laws, and by his selecting and combining power, to present a whole series of phenomena—such as ice frozen in contact with red-hot crucibles—which certainly did not belong to the "ordinary course of Nature." Such an exhibition a few centuries ago would, beyond all doubt, have subjected the lecturer on Heat to painful experience of that condition of matter. Nevertheless the phenomena so exhibited were natural phenomena—in this sense, that they were the product of natural laws. Only these laws were combined in action under extraordinary conditions, and these conditions were governed by the purpose and design of the lecturer, which design was "spontaneous," if there is any meaning in the word. In like manner, if the progress of discovery is as rapid during the next four hundred years as it has been during the last period of the same extent, men will be able to do many things which would now appear to be "supernatural." There is no difficulty in conceiving how a complete knowledge of all natural laws would give, if not complete power, at least degrees of power, immensely greater than those which we now possess. Power of this kind, then, however great in degree, clearly does not answer that idea of the Supernatural which so many reject as inconceivable. What, then, is that idea? Have we not traced it to its den at last? By "supernatural" power, do we not mean power independent of the use of means, as distinguished from power depending on knowledge—even infinite knowledge—of the means proper to be employed?

This is the sense—probably the only sense—in which the Supernatural is, to many minds, so difficult of belief. No man can have any difficulty in believing that there are natural laws of which he is ignorant; nor in conceiving that there may be Beings who do know them, and can use them, even as he himself now uses the few laws with which he is acquainted. The real difficulty lies in the idea of Will exercised without the use of means—not in the idea of Will exercised through means which are beyond our knowledge, or beyond our reach.

Now, have we any right to say that belief in *this* is essential to all Religion? If we have not, then, it is only putting, as so many other hasty sayings do put, additional difficulties in the way of Religion. The relation in which God stands to those rules of His government which are called "laws," is, of course, an inscrutable mystery to us. But the very idea of a Creator involves the idea not merely of a Being by whom the properties of Matter are employed, but of a Being from whose Will the properties of Matter are derived. This, indeed, is the proper work of Creation, as nearly as we can form a conception of it. It is true that in forming this conception we pass beyond the bounds of our own experience, because "we pass from that in God of which there is an image in Man, to that which is distinctive of God as God." But this we must do in forming any idea of a God at all. We must conceive the Creator as first giving existence to the means, and then using them for the accomplishment of ends. "We cannot conceive of the original relation of this Universe to God as that of an infinite multitude of laws to an infinite Mind, having (only) perfect knowledge of them, and using this knowledge in turning them to account, in accomplishing designs of infinite wisdom.

We cannot conceive of infinite wisdom thus, as it were, finding infinite resources already existing."[1] All this is true. But those who believe that God's Will does govern the world, must believe that ordinarily, at least, He does govern it by the choice and use of means,—which means were again pre-established by Himself. Nor have we any certain reason to believe that He ever acts otherwise. Extraordinary manifestations of His Will—signs and wonders—may be wrought, for aught we know, by similar instrumentality—only by the selection and use of laws of which Man knows and can know nothing, and which, if he did know, he could not employ.[2]

1 I am glad to be able to quote these passages from one of my earliest and most valued friends, the Rev. J. McLeod Campbell. They occur in an Introduction to a new edition of his work on the "Nature of the Atonement" (Macmillan and Co. 1867)—an Introduction marked by characteristic depth of thought and feeling.

2 This chapter, originally published as an article in the *Edinburgh Review* for Oct. 1862, has been referred to in the remarkable work of Mr. Lecky on "The Rise and Influence of Rationalism in Europe," (vol. i. ch. ii. p. 195 note,) as conveying "a notion of a miracle which would not differ *generically* from a human act, though it would still be strictly available for evidential purposes." I am quite satisfied with this definition of the result. Beyond the immediate purposes of benevolence, which were served by almost all the miracles of the New Testament, the only other purpose which is ever assigned to them is an "evidential purpose"—that is, a purpose that they might serve as signs of the presence of superhuman knowledge, and of the working of superhuman power. They were performed—in short—to assist faith, and not to confound reason.

Here, then, we come upon the question of miracles—how we understand them? what we would define them to be? The common idea of a miracle is, a suspension or violation of the laws of Nature. This is a definition which places the essence of a miracle in a particular method of operation. But there is another definition which passes over the question of method altogether, and dwells only on the agency by which, and the purpose for which, a wonderful work is wrought. "We would confine the word miracle," says Dr. M'Cosh,[1] "to those events which were wrought in our world as a sign or proof of God making a supernatural interposition, or a revelation to Man." The two most essential conditions in this view of a miracle, are that it is a work wrought by a Divine power for a Divine purpose, and is of a nature such as could not be wrought by merely human contrivance. This definition of a miracle does not necessarily exclude the idea of God working by the use of means, provided they are such means as are out of human reach. Indeed, in an important note (p. 149), Dr. M'Cosh explains that miracles are not to be considered "as against Nature" in any other sense than that in which "one natural agent may be against another—as water may counteract fire." This eminent writer

[1] "The Supernatural in relation to the Natural." By the Rev. James M'Cosh, LL.D. Macmillan, Cambridge, 1861.

has approached the subject by the right method, because he has addressed himself first to the solution of the one question which is an essential preliminary to all subsequent discussion:—"How much is contained in the Natural?" Not until this question is answered, can the Supernatural be defined. Yet the answer given by Dr. M'Cosh shows the inherent and the insuperable difficulty which attends the giving of any answer at all. "In this world," he says, "there is a set of objects and agencies which constitute a system or Cosmos which may have relations to regions beyond, *but is all the while a self-contained sphere, with a space around it—an Island so far separated from other lands*. This system we call Nature" (p. 101). This definition of the Natural is perhaps as accurate and as full as any that can be given. It assumes, however, that the boundaries of the Natural are known. But the essential difficulty of separating between the Natural and the Supernatural is this—that the boundaries of the Natural are *not* known—that we cannot trace the shores of this "island"—that even if we could see any distinct separation between them and the space around them, we have not explored the "island" itself completely, and therefore we cannot say of any agency working therein, that it comes from beyond the Sea. Mr. Mansel, in his "Essay on Miracles," adopts the word "superhuman" as the most accurate expression of his meaning. He says, "A superhuman authority needs to be substantiated by superhuman evidence; *and what is superhuman is miraculous.*"[1] It is important to observe that this definition does not necessarily involve the idea of a "violation of the laws of Nature." It does not involve the idea of the exercise of Will apart from the use of means. It does not imply any exception to the great law of causation. It does not involve, therefore, that idea which appears to many so difficult of conception. It simply supposes, without any attempt to fathom the relation in which God stands to His own "laws," that out of His infinite knowledge of these laws, or of His infinite power of making them the instruments of His Will, He may and He does use them for extraordinary indications of His presence.[2]

The reluctance to admit, as belonging to the domain of Nature, any special exertion of Divine power for

[1] "Aids to Faith," p. 35. In another passage (p. 21), Mr. Mansel says, that in respect to the great majority of the miracles recorded in Scripture, "the supernatural element appears . . . in the exercise of a personal power transcending the limits of man's will. They are not so much *supermaterial* as *superhuman.*"

[2] I agree with Mr. J. M. Campbell when he says, in the Introduction already quoted, "It appears to me that we do not know enough to say, as regards anything transcending our knowledge of Law, in which way we should view it—whether as belonging to the system of Law, but to a region of it out of our sight, or as outside of that system, and as having the same immediate relation to God which the system of Law ultimately has."—P. xxxv.

special purposes, stands really in very close relationship to the converse notion, that where the operation of natural causes can be clearly traced, there the exertion of Divine power and Will is rendered less certain and less convincing. This is the idea which lies at the root of Gibbon's famous chapters on the spread of Christianity. He labours to prove that it was due to natural causes. In proving this, he evidently thinks he is disposing of the notion that Christianity spread by Divine power; whereas he only succeeds in pointing out some of the means which were employed to effect a Divine purpose. In like manner, the preservation of the Jews as a distinct People during so many centuries of complete dispersion, is a fact standing nearly, if not absolutely, alone in the history of the world. It is at variance with all other experience of the laws which govern the amalgamation with each other of different families of the human race. The case of the Gipsies has been referred to as somewhat parallel. But the facts of this case are doubtful and obscure, and such of them as we know involve conditions altogether dissimilar in kind. It is not surprising, therefore, that the preservation of the Jews, partly from the relation in which it stands to the apparent fulfilment of Prophecy, and partly from the extraordinary nature of the fact itself, is tacitly assumed by many persons to come strictly within the category of miraculous events. Yet in itself it is nothing more than a striking illustration how a departure from the "ordinary course of nature" may be effected through the instrumentality of means which are natural and comprehensible. An extraordinary resisting power has been given to the Jewish People against those dissolving and disintegrating forces which have caused the disappearance of every other race placed under similar conditions. They have been torn from home and country, and removed, not in a body, but in scattered fragments, over the world. Yet they are as distinct from every other people now as they were in the days of Solomon. Nevertheless this resisting power, wonderful though it be, is the result of special laws, overruling those in ordinary operation. It has been effected by the use of means. Those means have been superhuman—they have been beyond human contrivance and arrangement. But they belong to the region of the Natural. They belong to it not the less, but all the more, because in their concatenation and arrangement they seem to indicate the purpose of a living Will seeking and effecting the fulfilment of its designs. This is the manner after which our own living wills in their little sphere effect their little objects. Is it difficult to believe that after the same manner also the Divine Will, of which ours is the image only, works and effects its purposes?

Our own experience shows that the universal Reign of Law is perfectly consistent with a power of making those laws subservient to design—even when the knowledge of them is but slight, and the power over them slighter still. How much more easy, how much more natural, to conceive that the same universality is compatible with the exercise of that Supreme Will before which all are known, and to which all are servants! What difficulty in this view remains in the idea of the Supernatural? Is it any other than the difficulty in believing in the existence of a Supreme Will—in a living God? If this be the belief of which M. Guizot speaks when he says that it is essential to religion, then his proposition is unquestionably true. In this sense the difficulty of believing in the Supernatural, and the difficulty of believing in pure Theism, is one and the same. But if he means that it is necessary to religion to believe in even the occasional "violation of law,"—if he means that without such belief, signs and wonders cease to be evidences of Divine power,—then he announces a proposition which cannot be sustained. There is nothing in Religion incompatible with the belief that all exercises of God's power, whether ordinary or extraordinary, are effected through the instrumentality of means—that is to say, by the instrumentality of natural laws brought out, as it were, and used for a Divine purpose. To believe in the existence of miracles, we must indeed believe in the Superhuman and in the Supermaterial. But both these are familiar facts in Nature. We must believe also in a Supreme Will and a Supreme Intelligence; but this our own Wills and our own Intelligence not only enable us to conceive of, but compel us to recognise in the whole laws and economy of Nature. Her whole aspect "answers intelligently to our intelligence—mind responding to mind as in a glass."[1] Once admit that there is a Being who—irrespective of any theory as to the relation in which the laws of Nature stand to His Will—has at least an infinite knowledge of those laws, and an infinite power of putting them to use—then miracles lose every element of inconceivability. In respect to the greatest and highest of all—that restoration of the breath of life which is not more mysterious than its original gift—there is no answer to the question which Paul asks, "Why should it be thought a thing incredible by you that God should raise the dead?"

This view of miracles is well expressed by Principal Tulloch :—

"The stoutest advocate of interference can mean nothing more than that the Supreme Will has so moved the hidden springs of nature that a new issue arises

1 "Beginning Life: Chapters for Young Men on Religion, Study, and Business. Chap. iii., The Supernatural." By John Tulloch, D.D. Principal of St. Mary's College, St. Andrew's. Edinburgh, 1860. P. 29.

on given circumstances. The ordinary issue is supplanted by a higher issue. The essential facts before us are a certain set of phenomena, and a Higher Will moving them. How moving them? is a question for human definition; but the answer to which does not and cannot affect the Divine meaning of the change. Yet when we reflect that this Higher Will is everywhere reason and wisdom, it seems a juster as well as a more comprehensive view to regard it as operating by subordination and evolution rather than by 'interference' or 'violation.' According to this view, the idea of Law is so far from being contravened by the Christian miracles, that it is taken up by them and made their very basis. They are the expression of a Higher Law, working out its wise ends among the lower and ordinary sequences of life and history. These ordinary sequences represent nature—nature, however, not as an immutable fate, but a plastic medium through which a Higher Voice and Will are ever addressing us, and which, therefore, may be wrought into new issues when the Voice has a new message, and the Will a special purpose for us." [1]

It is well worthy of remark, that Locke, who laid great stress on the Christian miracles, as attesting the authority of those who wrought them, declines, nevertheless, to adopt the common definition of that in

[1] "Beginning Life," &c. pp. 85, 86. By John Tulloch, D.D.

which miraculous agency consists. "A miracle then," he says,[1] "I take to be a sensible operation, which, being above the comprehension of the spectator and, *in his opinion,* contrary to the established course of nature, *is taken by him* to be Divine." And in reply to the objection, that this makes a miracle depend on the opinions or knowledge of the spectator, he points out that this objection cannot be avoided by any of the definitions commonly adopted; because "it being agreed that a miracle must be that which surpasses the force of nature in the established steady laws of cause and effect, nothing can be taken to be a miracle but *what is judged to exceed those laws.* Now every one being able to judge of those laws *only by his own acquaintance with nature,* and his own notions of its force, which are different in different men, it is unavoidable that that should be a miracle to one man which is not so to another." In this passage Locke recognises the great truth, that we can never know what is above Nature unless we know all that is within Nature. But he misses another truth, quite as important,—that a miracle would still be a miracle even though we did know the laws through which it was accomplished, provided those laws, though not beyond human knowledge, were beyond human control. We might know the conditions necessary to the performance

[1] "A Discourse on Miracles."

of a miracle, although utterly unable to bring those conditions about. Yet a work performed by the bringing about of conditions which are out of human reach, would certainly be a work attesting superhuman power.

Nevertheless so deeply ingrained in popular theology is the idea that miracles, to be miracles at all, must be performed by some violation or suspension of the laws of Nature, that the opposite idea of miracles being performed by the use of means is regarded by many with jealousy and suspicion. Strange that it should be thought the safest course to separate as sharply and as widely as we can between what we are called upon to believe in Religion, and what we are able to trace or understand in Nature! With what heart can those who cherish this frame of mind follow the great argument of Butler? All the steps of that argument—the greatest in the whole range of Christian philosophy—are founded on the opposite belief, that all the truths, and not less all the difficulties of Religion, have their type and likeness in the "constitution and course of Nature." As we follow that reasoning, so simple and so profound, we find our eyes ever opening to some new interpretation of familiar facts, and recognising among the curious things of earth, one after another of the laws which, when told us of the spiritual world, seem so perplexing and so hard to accept or understand. To ask how much further this argument of the "Analogy" is capable of illustration and development, is to ask how much more we shall know of Nature. Like all central truths, its ramifications are infinite—as infinite as the appearance of variety, and as pervading as the sense of oneness in the universe of God.

But what of Revelation? Are its history and doctrines incompatible with the belief that God uniformly acts through the use of means? The narrative of Creation is given to us in abstract only, and is told in two different forms, both having apparently for their main, perhaps their exclusive object, the presenting to our conception the personal agency of a living God. Yet this narrative indicates, however slightly, that room is left for the idea of a material process. "Out of the dust of the ground;" that is, out of the ordinary elements of Nature, was that Body formed which is still upheld and perpetuated by organic forces acting under the rules of Law. Nothing which Science has discovered, or can discover, is capable of traversing that simple narrative. On this subject M. Guizot lays great stress, as many others do, on what he calls the Supernatural in Creation, as distinguished from the operations now visible in Nature. "De quelle façon et par quelle puissance le genre humain a-t-il commencé sur la terre?" In reply to this question, he proceeds to argue that Man must have been the result either of mere material forces, or of a supernatural power exterior to, and superior to Matter. Spontaneous generation, he argues, supposing it to exist at all, can give birth only to infant beings—to the first hours, and feeblest forms of nascent life. But Man—the human pair—must evidently have been complete from the first; created in the full possession of their powers and faculties. "C'est à cette condition seulement qu'en apparaissant pour la première fois sur la terre l'homme aurait pu y vivre—s'y perpétuer, et y fonder le genre humain. Evidemment l'autre origine du genre humain est seul admissible, seul possible. Le fait surnaturel de la création explique seul la première apparition de l'homme ici-bas."*

This is a common but not a very safe argument. If the Supernatural—that is to say, the Superhuman and the Supermaterial—cannot be found nearer to us than this, it will not be securely found at all. It is very difficult to free ourselves from this notion that by going far enough back, we can "find out God" in some sense in which we cannot find Him now. The certainty not merely of one, but of many successive Creations in the history of our Planet, and especially of a time comparatively recent, when Man did not exist, is indeed an effectual answer to the notion, if it be now ever entertained, of "all things having continued as they are since the Beginning." But those who believe that the existing processes of Nature can be accounted for by "Law," may believe that those processes were also commenced by the same vague and mysterious agency. To accept the primeval narrative of the Jewish Scriptures as coming from authority, and as bringing before us the personal agency of the Creator, but without purporting to reveal the method of His work,—this is one thing. To argue that no other origin for the first parents of the human race is conceivable than that they were moulded perfect, without the instrumentality of any means,—this is quite another thing. The various hypotheses of Development, of which Darwin's theory is only a new and special version, whether they are probable or not, are at least advanced as affording a possible escape from the puzzle which M. Guizot puts. These hypotheses are indeed destitute of proof; and in the form which they have as yet assumed, it may justly be said that they involve such violations of, or departures from, all that we know of the existing order of things, as to deprive them of all scientific basis. But the close and mysterious relations between the mere animal frame of Man, and that of the lower animals, does render the idea of a common relationship by descent at least conceivable. Indeed, in proportion as it seems to approach nearer to processes of which we have some knowledge, it is, in a degree, more conceivable than Creation without any process,—of which we have no knowledge and can have no conception.

But whatever may have been the method or process of Creation, it is Creation still. If it were proved to-morrow that the first man was "born" from some pre-existing Form of Life, it would still be true that such a birth must have been, in every sense of the word, a new Creation. It would still be as true that God formed him "out of the dust of the earth," as it is true that He has so formed every child who is now called to answer the first question of all theologies. And we must remember that the language of Scripture nowhere draws, or seems even conscious of, the distinction which modern philosophy draws so sharply between the Natural and the Supernatural. All the operations of Nature are spoken of as operations of the Divine Mind. Creation is the outward embodiment of a Divine idea. It is in this sense, apparently, that the narrative of Genesis speaks of every plant being formed "before it grew." But the same language is held, not less decidedly, of every ordinary birth. "Thine eyes did see my substance, yet being imperfect. In Thy book all my members were written, which in continuance were fashioned, when as yet there were none of them." And these words, spoken of the individual birth, have been applied not less truly to the modern idea of the Genesis of all Organic Life. Whatever may have been the physical or material relation between its successive forms, the ideal relation has been now clearly recognised, and reduced to scientific definition. All the members of that frame which has received its highest interpretation in Man, had existed, with lower offices assigned to them, in the animals which flourished before Man was born. All theories of Development have been simply attempts to suggest the manner in which, or the physical process by means of which, this ideal continuity of type and pattern has been preserved. But whilst all these suggestions have been in the highest degree uncertain, some of them violently absurd, the one thing which is certain is the fact for which they endeavour to account. And what is that fact? It is one which belongs to the world of Mind, not to the world of Matter. When Professor Owen tells us, for example, that certain jointed bones in the Whale's paddle are the same bones which in the Mole enable it to burrow, which in the Bat enable it to fly, and in Man constitute his hand with all its wealth of functions, he does not mean that physically and actually they are the same bones, nor that they have the same uses, nor that they ever have been, or ever can be, transferable from one kind of animal to another. He means that in a purely ideal or mental conception of the plan of all Vertebrate skeletons, these bones occupy the same relative place—relative, that is, not to origin or use, but to the Plan or conception of that skeleton as a whole.

Here the Supermaterial, and in this sense the Supernatural, element,—that is to say, the ideal conformity and unity of conception, is the one unquestionable fact, in which we recognise directly the working of a Mind with which our own has very near relations. Here, as elsewhere, we see the Natural, in the largest sense, including and embodying the Supernatural; the Material, including the Supermaterial. No possible theory, whether true or false, in respect to the physical means employed to preserve the correspondence of parts which runs through all Creation, can affect the certainty of that mental plan and purpose which alone makes such correspondence intelligible to us, and in which alone it may be said to exist.

It must always be remembered that the two ideas,—that of a Physical Cause and that of a Mental Purpose,—are not antagonistic; only the one is larger and more comprehensive than the other. Let us take a case. In many animal frames there are what have been called "silent members"—members which have no reference to the life or use of the animal, but only to the general pattern on which all vertebrate skeletons have been formed. Mr. Darwin, when he sees such a member in any animal, concludes with certainty that this animal is the lineal descendant by ordinary generation of some other animal in which that member was not silent but turned to use. Professor Owen, taking a larger and wider view, would say, without pretending to explain *how* its presence is to be accounted for physically, that the silent member has relation to a general purpose or plan which can be traced from the dawn of Life, but which did not receive its full accomplishment until Man was born. This is certain: the other is a theory. The assumed physical cause may be true or false. But in any case the mental purpose and design—the conformity to an abstract idea—this is certain. The relation in which created Forms stand to our own mind and to our understanding of their Purpose, is the one thing which we can surely know, because it belongs to our own consciousness. It is entirely independent of any belief we may entertain, or any knowledge we may acquire, of the processes employed for the fulfilment of that Purpose.

And yet scientific men sometimes tell us that "we must be very cautious how we ascribe intention to Nature. Things do fit into each other, no doubt, as if they were designed; but all we know about them is that these correspondences exist, and that they seem to be the result of physical laws of development and growth." Very likely; but how these correspondences have arisen, and are daily arising, is not the question, and it is immaterial how that question may be answered. Do those correspondences exist, or do they not? The perception of them by our mind is as much a fact as the sight or touch of the things in which they appear. They may have been produced by growth—they may have been the result of a process of development,—but it is not the less the development of a mental purpose. It is the end subserved that we absolutely know. What alone is doubtful and obscure is precisely that which we are told is the only legitimate object of our research,—viz., the means by which that end has been attained. Take one instance out of millions. The poison of a deadly snake —let us for a moment consider what this is. It is a secretion of definite chemical properties which have reference, not only—not even mainly—to the organism of the animal in which it is developed, but specially to the organism of another animal which it is intended to destroy. Some naturalists have a vague sort of notion that, as regards merely mechanical weapons, or organs of attack, they may be developed by use,—that legs may become longer by fast running, teeth sharper and longer by much biting. Be it so: this law of growth, if it exist, is but itself an instrument whereby purpose is fulfilled. But how will this law of growth adjust a poison in one animal with such subtle knowledge of the organisation of another that the deadly virus shall in a few minutes curdle the blood, benumb the nerves, and rush in upon the citadel of life? There is but one explanation—a Mind, having minute and perfect knowledge of the structure of both, has designed the one to be capable of inflicting death upon the other. This mental purpose and resolve is the one thing which our intelligence perceives with direct and intuitive recognition. The method of creation, by means of which this purpose has been carried into effect, is utterly unknown.

It is no answer or objection to this view that poisons exist also in plants and minerals where no similar adjustment to function is perceived.[1] Even in these cases there are wonderful relations between our own human frame and many poisons of the mineral and vegetable world which render them invaluable agents in the mitigation of suffering and the prevention or removal of disease. It is impossible to believe that these complicated relations of action and reaction between things separated apparently from each other by the whole width of being, have been the result of forces with which Mind and Prevision have had no concern. But even if the use of such poisons were absolutely unknown—even if that use lay, which it does not, beyond the possibility of our conception,—this would not deduct by the value of a fraction from the certainty of a conclusion which is founded on different conditions. The relations of adjustment between a given number of elements are none the less a certain fact because similar elements may be found elsewhere without any such adjustment being visible to us. It is the very fact of their not being

[1] "To what intention are we to ascribe the poisons liberally distributed through plants and minerals?" asks Mr. G. H. Lewes in his review of this work.—*Fortnightly Review*, July 1867, p. 100.

separate but combined in the one case which justifies and compels a conclusion different from that which arises in the other case. This is the law of evidence on which we act and judge in other matters with conviction which is both intuitive, and capable of being confirmed by the rules of reason. And this reply is applicable to all objections of the same kind. Those portions of the system of Nature which are wholly dark to us do not necessarily cast any shadow on those other portions of that system which are luminous with inherent light. Rather the other way. The shining tracts which thus reflect the light of Reason and of Mind send abundant rays into all the dark places round them. The new discoveries which Science is ever making of adjustments and combinations, of which we had no previous conception, impress us with an irresistible conviction that the same relations to Mind prevail throughout. It matters not in what department of investigation inquiry is conducted, it matters not what may be the Philosophy or Theology of the inquirer. Every step he takes he finds himself face to face with facts which he cannot describe intelligibly either to himself or others, except by referring them to that function and power of Mind which we know as Purpose and Design.

Perhaps no illustration more striking of this principle

was ever presented than in the curious volume published by Mr. Darwin on the "Fertilisation of Orchids."[1] It appears that the fertilisation of almost all Orchids is dependent on the transport of the pollen from one flower to another by means of insects. It appears, further, that the structure of these flowers is elaborately contrived, so as to secure the certainty and effectiveness of this operation. Mr. Darwin's work is devoted to tracing in detail what these contrivances are. To a large extent they are purely mechanical, and can be traced with as much clearness and certainty as the different parts of which a steam-engine is composed. The complication and ingenuity of these contrivances almost exceed belief. "Moth-traps and spring-guns set on these grounds," might be the motto of the Orchids. There are baits to tempt the nectar-loving Lepidoptera, with rich odours exhaled at night, and lustrous colours to shine by day; there are channels of approach along which they are surely guided, so as to compel them to pass by certain spots; there are adhesive plasters nicely adjusted to fit their probosces, or to catch their brows; there are hair triggers carefully set in their necessary path, communicating with explosive shells, which project the pollen-

1 "On the various Contrivances by which British and Foreign Orchids are fertilised by Insects." By Charles Darwin, F.R S. London, 1862.

stalks with unerring aim upon their bodies. There are, in short, an infinitude of adjustments, for an idea of which I must refer my readers to Mr. Darwin's inimitable powers of observation and description—adjustments all contrived so as to secure the accurate conveyance of the pollen of the one flower to its precise destination in the structure of another.

Now there are two questions which present themselves when we examine such a mechanism as this. The first is, What is the use of the various parts, or their relation to each other with reference to the purpose of the whole? The second question is, How were those parts made, and out of what materials? It is the first of these questions—that is to say, the use, object, intention, or purpose of the different parts of the plant—which Darwin sets himself instinctively to answer first; and it is this which he does answer with precision and success. The second question,—that is to say, how those parts came to be developed, and out of what "primordial elements" they have been derived in their present shapes, and converted to their present uses—this is a question which Darwin does also attempt to solve, but the solution of which is in the highest degree difficult and uncertain. It is curious to observe the language which this most advanced disciple of pure naturalism instinctively uses when he has to describe the complicated structure of this curious order of plants. "Caution in ascribing intentions to nature," does not seem to occur to him as possible. Intention is the one thing which he does see, and which, when he does not see, he seeks for diligently until he finds it. He exhausts every form of words and of illustration by which intention or mental purpose can be described. "Contrivance"—"curious contrivance"—"beautiful contrivance,"—these are expressions which recur over and over again. Here is one sentence describing the parts of a particular species: "the Labellum is developed into a long nectary, *in order* to attract Lepidoptera, and we shall presently give reasons for suspecting that the nectar is *purposely* so lodged that it can be sucked only slowly, *in order* to give time for the curious chemical quality of the viscid matter setting hard and dry."[1] Nor are these words used in any sense different from that in which they are applicable to the works of Man's contrivance—to the instruments we use or invent for carrying into effect our own preconceived designs. On the contrary, human instruments are often selected as the aptest illustrations both of the object in view, and of the means taken to effect it. Of one particular structure, Mr. Darwin says: "This contrivance of the guiding ridges may be compared to the little instrument sometimes

[1] P. 29.

used for guiding a thread into the eye of a needle." Again, referring to the precautions taken to compel the insects to come to the proper spot, in order to have the "pollinia" attached to their bodies, Mr. Darwin says: "Thus we have the rostellum partially closing the mouth of the nectary, *like a trap placed in a run for game*,—and the trap so complex and perfect!"[1] But this is not all. The idea of special use, as the controlling principle of construction, is so impressed on Mr. Darwin's mind, that, in every detail of structure, however singular or obscure, he has absolute faith that in this lies the ultimate explanation. If an organ is largely developed, it is because some special purpose is to be fulfilled. If it is aborted or rudimentary, it is because that purpose is no longer to be subserved. In the case of another species whose structure is very singular, Mr. Darwin had great difficulty in discovering how the mechanism was meant to work, so as to effect the purpose. At last he made it out, and of the clue which led to the discovery he says: "The strange position of the Labellum perched on the summit of the column, ought to have shown me that here was the place for experiment. I ought to have scorned the notion that the Labellum was thus placed *for no good purpose*. I neglected this plain guide, and for a long time completely failed to understand the flower."[2]

1 P. 30. 2 P. 262.

An attempt has, indeed, been made to explain away Mr. Darwin's language in such cases as "metaphorical."[1] But this explanation is powerless to expel from that language the inference it involves. Indeed, it is an explanation which only repeats the same idea in another form. The very essence of a metaphor is that it expresses the resemblances of things. But it is in seeing the resemblances, and in seeing the correlative differences of things, that all knowledge consists. This perception is the raw material of Thought—it is the foundation of all intellectual apprehension. In proportion as resemblances are complete, the language which expresses those resemblances is the language of truth. Such language very often carries within it the most certain conclusions which are accessible to reason. One mind looking at the workings of another mind can see likeness of agency only by recognising likeness in the processes of thought. That likeness can only be expressed in words which convey the idea of it to other minds. But in this sense all language is metaphorical. The commonest words we use to indicate ideas are essentially metaphorical, bringing home into the world of Mind images

[1] *Quarterly Journal of Science*, Oct. 1867. "Creation by Law," by Alfred Wallace. "Mr. Darwin has laid himself open to much misconception, and has given to his opponents a powerful weapon by his continual use of metaphor in describing the wonderful co-adaptations of organic beings."—P. 473.

derived from material force, and carrying forth again into the outward world conceptions born of that mental power which alone is capable of conceiving. In one aspect, all human speech is what the Poet calls it, "Matter-moulded forms of speech."[1] In another aspect it is all spirit-moulded, since we can only think of Matter in the light of those impressions which it has power to make on Mind. All language is thus but a system of signs whereby we express the analogies—the differences and resemblances perceived by us in those two great departments of Nature of which the union and the separation are both imaged in ourselves—that is, in the union and in the difference of the Body and the Mind. The most absolute certainties we can ever know are only known by the translation of ideas or conceptions from one of these departments to the other, and the language in which these certainties are expressed carries, and must carry, signs of this origin in itself. The question, therefore, in respect to Mr. Darwin's language, is not whether it is "metaphorical"—that is, whether it applies to material phenomena conceptions derived from the world of Mind. This, of course, it does, and in the nature of the case it must do. But the question is, whether the correspondence it expresses between the order of these material phenomena and a known order

1 "In Memoriam," xciv.

of Thought is or is not a real correspondence, and one, therefore, indicating the known effects of a known originating cause.

And here it is well worthy of observation, that although Purpose and Intention are, of course, involved in all mental operations, yet the conception of contrivance is not the only mental conception which, in like manner, is recognised as constituting the order of natural phenomena. Other conceptions equally familiar to the mind of Man are instinctively recognised by all Naturalists who bring high intellectual powers into that contact with Nature which consists in close and thoughtful observation of her facts. Other mental conceptions, such as those of Number and Proportion, are then found to emerge, and make an ineffaceable impression on the mind which sees them.

Thus, when we come to the second part of Mr. Darwin's work, viz. the Homology of the Orchids, we find that the inquiry divides itself into two separate questions,—first, the question what all these complicated organs are in their primitive relation to each other; and, secondly, how these successive modifications have arisen, so as to fit them for new and changing uses. Now, it is very remarkable that of these two questions, that which may be called the most abstract and transcendental—the most nearly related to the Supernatural and the Supermaterial—is again precisely the one which Darwin is able to solve most clearly. We have already seen how well he solves the first question—What is the use and intention of these various parts? The next question is, What are these parts in their primal order and conception? The answer is, that they are members of a numerical group, having a definite and still traceable order of symmetrical arrangement. They are expressions of a numerical idea, as so many other things—perhaps as all things—of beauty are. Mr. Darwin gives a diagram, showing the primordial or archetypal arrangement of Threes within Threes, out of which all the strange and marvellous forms of the Orchids have been developed, and to which, by careful counting and dissection, they can still be ideally reduced. But when we come to the last question—By what process of natural consequence have these elementary organs of Three within Three been developed into so many various forms of beauty, and made to subserve so many curious and ingenious designs?—we find nothing but the vaguest and most unsatisfactory conjectures. Let us take one instance as an example. There is a Madagascar Orchis—the "Angræcum sesquipedale"—with an immensely long and deep nectary. How did such an extraordinary organ come to be developed? Mr. Darwin's explanation is this: The pollen of this flower can only be removed by the proboscis of some very large Moth trying to get at the nectar at the bottom of the vessel. The Moths with the longest probosces would do this most effectually; they would be rewarded for their long noses by getting the most nectar; whilst, on the other hand, the flowers with the deepest nectaries would be the best fertilised by the largest Moths preferring them. Consequently, the deepest-nectaried Orchids, and the longest-nosed Moths, would each confer on the other a great advantage in the "battle of life." This would tend to their respective perpetuation, and to the constant lengthening of nectaries and of noses. But the passage is so curious and characteristic, that it is well to give Mr. Darwin's own words:—

"As certain Moths of Madagascar became larger, through natural selection in relation to their general conditions of life, either in the larval or mature state, or as the proboscis alone was lengthened to obtain honey from the Angræcum, those individual plants of the Angræcum which had the longest nectaries, (and the nectary varies much in length in some Orchids,) and which, consequently, compelled the Moths to insert their probosces up to the very base, would be the best fertilised. These plants would yield most seed, and the seedlings would generally inherit longer nectaries; and so it would be in successive generations of the plant and

Moth. Thus it would appear that there has been a race in gaining length between the nectary of the Angræcum and the proboscis of certain Moths; but the Angræcum has triumphed, for it flourishes and abounds in the forests of Madagascar, and still troubles each Moth to insert its proboscis as far as possible in order to drain the last drop of nectar. . . . . We can thus," says Mr. Darwin, "*partially* understand how the astonishing length of the nectary may have been acquired by successive modifications."

It is indeed but a "partial" understanding.[1] How came this Orchis to require any exact adjustment between the length of its nectary and the proboscis of an insect? This is not a general necessity even among the Orchids. "In the British species, such as Orchis Pyramidalis, it is not necessary that any such adjustment should exist, and thus a number of insects of various sizes are found to carry away the pollinia, and aid in the fertilisation."[2] This would obviously be the most favourable condition for all Orchids in the battle of life. Does not the hypothesis, then, begin by assuming the very

[1] The passage which follows I have added to meet the objection taken by Mr. Wallace, that I have "not shown what point the explanation fails to meet." A sample only of such points can be given here. See also Note A.

[2] "Creation by Law." G. A. R. Wallace. *Journal of Science*, October 1867, p. 475.

condition of things for which it professes to account? We must start with this Madagascar Orchis already in possession of a larger nectary than other species, and with a structure already depending on particular Moths also already existing, and already provided with probosces of nicely adjusted length. If the nectaries began first to lengthen, how came the Moths not to leave them for other flowers? And if, on the contrary, they began to shorten, how came they not to be favoured and resorted to by other Moths of a smaller size? Can we assume that somehow there were always ready some Moths still larger to favour the longer variety, and that somehow also there were no smaller Moths to favour the shorter?[1] Why should the race *in this particular species* be always in the direction of nectaries getting longer, and not rather in the direction of nectaries getting shorter? Obviously the same hypothesis might be so turned as to account for either result with equal ease, and therefore it does not account at all for one of those results as against the other. And then there is

[1] Mr. Wallace sees no difficulty whatever in making any supposition of this kind which the Theory may require. "Now let us start," he says, "from the time when the nectary was only half its present length, or about six inches, and was chiefly fertilized by a species of Moth *which appeared at the time of the plants flowering, and whose proboscis was of the same length*."—Ibid, p. 475.

a larger question than any of these which remains behind. How came Orchids to be dependent at all upon insects for fertilisation? It cannot be argued that this is a necessity arising mechanically from the nature of things, because, as we are truly told by an eminent naturalist who warmly supports the Darwinian hypothesis, "exactly the same end is attained in ten thousand other flowers" which do not possess the same structure.[1] But what is the bearing of this fact upon the theory? Is it not this—that the origin of such curious structures, and complicated relations, cannot be accounted for on any principle of mere mechanical necessity? Elementary forces may indeed always be detected, for they are always present. But the manner in which they are worked irresistibly suggests some directing power, having as one of its aims mere increase and variety in that ocean of enjoyment which constitutes the sum of Organic Life. Some idea of this kind, however unconsciously, however reluctantly conceded, lurks in every form of words in which the facts of science can be generalised to the mind. Thus we find Mr. Wallace himself saying, in the same paper in which he regrets the language of Mr. Darwin, that the conception he prefers is, that the "contrivances" referred to "are some of the results of those general laws which

[1] "Creation by Law," p. 474.

were *so co-ordinated* at the first introduction of Life upon the earth, as to result necessarily in the utmost possible development of *varied* forms." Eliminating the word "necessarily," which, if it has any meaning, does not apply, as we have seen, to the case of the Orchids, this language presents an intelligible idea. It satisfies the mind precisely in proportion as it brings into view, however distant, the attributes of Mind, and gives us a glimpse of "the reason why." The production of variety in beauty and in enjoyment is the purpose which those words suggest. In like proportion is Mr. Darwin's language the truest and the best. His explanations of the mechanical methods by which a wonderful Orchis has come to be are indeed, as he himself says, with great candour, "partial" and partial only. How different from the clearness and the certainty with which Mr. Darwin is able to explain to us the use and intention of the various organs! or the primal idea of numerical order and arrangement which governs the whole structure of the flower! It is the same through all Nature. Purpose and intention, or ideas of order based on numerical relations, are what meet us at every turn, and are more or less readily recognised by our own intelligence as corresponding to conceptions familiar to our own minds. We know, too, that these purposes and ideas are not our own, but the ideas and purposes of Another—of One whose manifestations are indeed superhuman and supermaterial, but are not "supernatural," in the sense of being strange to Nature, or in violation of it.

The truth is, that there is no such distinction between what we find in Nature, and what we are called upon to believe in Religion, as that which men pretend to draw between the Natural and the Supernatural. It is a distinction purely artificial, arbitrary, unreal. Nature presents to our intelligence, the more clearly the more we search her, the designs, ideas, and intentions of some

"Living Will that shall endure,
When all that seems shall suffer shock."

Religion presents to us that same Will, not only working equally through the use of means, but using means which are strictly analogous—referable to the same general principles—and which are constantly appealed to as of a sort which we ought to be able to appreciate, because we ourselves are already familiar with the like. Religion makes no call on us to reject that idea, which is the only idea some men can see in Nature—the idea of the universal Reign of Law—the necessity of conforming to it—the limitations which in one aspect it seems to place on the exercise of Will,—the essential basis, in another aspect, which it supplies for all the functions of Volition. On the contrary, the high regions into which this idea is found extending, and the matters over which it is found prevailing, is one of the deepest mysteries both of Religion and of Nature. We feel sometimes as if we should like to get above this rule—into some secret Presence where its bonds are broken. But no glimpse is ever given us of anything, but "Freedom within the bounds of Law." The Will revealed to us in Religion is not—any more than the Will revealed to us in Nature—a capricious Will, but one with which, in this respect, "there is no variableness, neither shadow of turning."

We return, then, to the point from which we started. M. Guizot's affirmation that belief in the Supernatural is essential to all Religion is true only when it is understood in a special sense. Belief in the existence of a Living Will—of a Personal God—is indeed a requisite condition. Conviction "that He is" must precede the conviction that "He is the rewarder of those that diligently seek Him." But the intellectual yoke involved in the common idea of the Supernatural is a yoke which men impose upon themselves. Obscure thought and confused language are the main source of difficulty.

Assuredly, whatever may be the difficulties of Christianity, *this* is not one of them,—that it calls on us to believe in any exception to the universal prevalence and power of Law. Its leading facts and doctrines are directly connected with this belief, and directly suggestive of it. The Divine mission of Christ on earth—

does not this imply not only the use of means to an end, but some inscrutable necessity that certain means, and these only, should be employed in resisting and overcoming evil? What else is the import of so many passages of Scripture implying that certain conditions were required to bring the Saviour of Man into a given relation with the race He was sent to save? "It behoved Him . . . . to make the Captain of our Salvation perfect through suffering." "It behoved Him in all things to be made like unto His brethren, *that He might be*," &c.—with the reason added: "for *in that* He Himself hath suffered being tempted, *He is able* to succour them that are tempted." Whatever *more* there may be in such passages, they all imply the universal reign of Law in the moral and spiritual, as well as in the material world: that those laws had to be—behoved to be—obeyed; and that the results to be obtained are brought about by the adaptation of means to an end, or, as it were, by way of natural consequence from the instrumentality employed. This, however, is an idea which systematic theology generally regards with intense suspicion, though, in fact, all theologies involve it, and build upon it. But then they are very apt to give explanations of that instrumentality which have no counterpart in the material or in the moral world. Perhaps it is not too much to say that the manifest decay which so many creeds and confessions are now suffering, arises mainly from the degree in which at least the popular expositions of them dissociate the doctrines of Christianity from the analogy and course of Nature. There is no such severance in Scripture—no shyness of illustrating Divine things by reference to the Natural. On the contrary, we are perpetually reminded that the laws of the spiritual world are in the highest sense laws of Nature, whose obligation, operation, and effect are all in the constitution and course of things. Hence it is that so much was capable of being conveyed in the form of parable—the common actions and occurrences of daily life being often chosen as the best vehicle and illustration of the highest spiritual truths. It is not merely, as Jeremy Taylor says, that "all things are full of such resemblances,"—it is more than this—more than resemblance. It is the perpetual recurrence, under infinite varieties of application, of the same rules and principles of Divine government,—of the same Divine thoughts, Divine purposes, Divine affections. Hence it is that no verbal definitions or logical forms can convey religious truth with the fulness or accuracy which belongs to narratives taken from Nature—Man's nature and life being, of course, included in the term:

> "And so, the Word had breath, and wrought
> With human hands the Creed of creeds."[1]

[1] Tennyson's "In Memoriam."

The same idea is expressed in the passionate exclamation of Edward Irving:—"We must speak in parables, or we must present a wry and deceptive form of truth; of which choice the first is to be preferred, and our Lord adopted it. Because parable is truth veiled, not truth dismembered; and as the eye of the understanding grows more piercing, the veil is seen through, and the truth stands revealed." Nature is the great Parable; and the truths which she holds within her are veiled, but not dismembered. The pretended separation between that which lies within Nature and that which lies beyond Nature is a dismemberment of the truth. Let both those who find it difficult to believe in anything which is "above" the Natural, and those who insist on that belief, first determine how far the Natural extends. Perhaps in going round these marches they will find themselves meeting upon common ground. For, indeed, long before we have searched out all that the Natural includes, there will remain little in the so-called Supernatural which can seem hard of acceptance or belief—nothing which is not rather essential to our understanding of this otherwise "unintelligible world"

CHAPTER II.

LAW;—ITS DEFINITIONS.

THE Reign of Law—is this, then, the reign under which we live? Yes, in a sense it is. There is no denying it. The whole world around us, and the whole world within us, are ruled by Law. Our very spirits are subject to it—those spirits which yet seem so spiritual, so subtle, so free. How often in the darkness do they feel the restraining walls—bounds within which they move—conditions out of which they cannot think! The perception of this is growing in the consciousness of men. It grows with the growth of knowledge; it is the delight, the reward, the goal of Science. From Science it passes into every domain of thought, and invades, amongst others, the Theology of the Church. And so we see the men of Theology coming out to parley with the men of Science,—a white flag in their hands, and saying, "If you will let us alone we will do the same by you. Keep to your own province, do not enter ours. The Reign of Law which you proclaim, we admit—outside these walls, but not within them:—let there be peace between us." But this will never do. There can be no such treaty dividing the domain of Truth. Every one Truth is connected with every other Truth in this great Universe of God. The connexion may be one of infinite subtlety, and apparent distance—running, as it were, underground for a long way, but always asserting itself at last, somewhere, and at some time. No bargaining, no fencing off the ground—no form of process, will avail to bar this right of way. Blessed right, enforced by blessed power! Every truth, which is truth indeed, is charged with its own consequences, its own analogies, its own suggestions. These will not be kept outside any artificial boundary; they will range over the whole Field of Thought, nor is there any corner of it from which they can be warned away.

And therefore we must cast a sharp eye indeed on every form of words which professes to represent a scientific truth. If it be really true in one department of thought, the chances are that it will have its bearing on every other. And if it be not true, but erroneous, its effect will be of a corresponding character; for there is a brotherhood of Error as close as the brotherhood of Truth. Therefore, to accept as a truth that which is not a truth, or to fail in distinguishing the sense in which a proposition may be true, from other senses in which it is not true, is an evil having consequences which are indeed incalculable. There are subjects on which one mistake of this kind will poison all the wells of truth, and affect with fatal error the whole circle of our thoughts.

It is against this danger that some men would erect a feeble barrier by defending the position, that Science and Religion may be, and ought to be, kept entirely separate ;—that they belong to wholly different spheres of thought, and that the ideas which prevail in the one province have no relation to those which prevail in the other. This is a doctrine offering many temptations to many minds. It is grateful to scientific men who are afraid of being thought hostile to Religion. It is grateful to religious men who are afraid of being thought to be afraid of Science. To these, and to all who are troubled to reconcile what they have been taught to believe with what they have come to know, this doctrine affords a natural and convenient escape. There is but one objection to it—but that is the fatal objection—that it is not true. The spiritual world and the intellectual world are not separated after this fashion : and the notion that they are so separated does but encourage men to accept in each, ideas which will at last be found to be false in both. The truth is, that there is no branch of human inquiry, however purely physical, which is more than the word "branch" implies;—none which is not connected through endless ramifications with every other,—and especially that which is the root and centre of them all. If He who formed the mind be one with Him who is the Orderer of all things concerning which that mind is occupied, there can be no end to the points of contact between our different conceptions of them, of Him, and of ourselves.

The instinct which impels us to seek for harmony in the truths of Science and the truths of Religion, is a higher instinct and a truer one than the disposition which leads us to evade the difficulty by pretending that there is no relation between them. For, after all, it is a pretence and nothing more. No man who thoroughly accepts a principle in the philosophy of Nature which he feels to be inconsistent with a doctrine of Religion, can help having his belief in that doctrine shaken and undermined. We may believe, and we must believe, both in Nature and Religion, many things which we cannot understand; but we cannot really believe two propositions which are felt to be contradictory. It helps us nothing in such a difficulty, to say that the one proposition belongs to Reason and the other proposition belongs to Faith. The endeavour to reconcile them is a necessity of the mind. We are right in thinking that, if they are both indeed true, they can be reconciled, and if they really are fundamentally opposed, they cannot both be true. That is to say, there must be some error in our manner of conception in one or in the other, or in both. At the very best, each can represent only some partial and imperfect aspect of the truth. The error may lie in our Theology, or it may lie in what we are pleased to call our Science. It may be that some dogma, derived by tradition from our fathers, is having its hollowness betrayed by that light which sometimes shines upon the ways of God out of a better knowledge of His works. It may be that some proud and rash generalisation of the schools is having its falsehood proved by the violence it does to the deepest instincts of our spiritual nature,—to

> "Truths which wake to perish never!
> Which neither man nor boy,
> Nor all that is at enmity with joy,
> Can utterly abolish or destroy."[1]

Such, for example, is the conclusion to which the language of some scientific men is evidently pointing, that great general Laws inexorable in their operation,

1 "Ode to Immortality, from the Recollections of early Childhood."—Wordsworth.

and Causes in endless chain of invariable sequence, are the governing powers in Nature, and that they leave no room for any special direction or providential ordering of events. If this be true, it is vain to deny its bearing on Religion. What then can be the use of prayer? Can Laws hear us? Can they change, or can they suspend themselves? These questions cannot but arise, and they require an answer. It is said of a late eminent Professor and clergyman of the English Church, who was deeply imbued with these opinions on the place occupied by Law in the economy of Nature, that he went on, nevertheless, preaching high doctrinal sermons from the pulpit until his death. He did so on the ground that propositions which were contrary to his reason were not necessarily beyond his faith. The inconsistencies of the human mind are indeed unfathomable; and there are men so constituted as honestly to suppose that they can divide themselves into two spiritual beings, one of whom is sceptical, and the other is believing. But such men are rare—happily for Religion, and not less happily for Science. No healthy intellect, no earnest spirit, can rest in such self-betrayal. Accordingly we find many men now facing the consequences to which they have given their intellectual assent, and taking their stand upon the ground that prayer to God has no other value or effect than so far as it may be a good way of preaching to ourselves. It is a useful and helpful exercise for our own spirits, but it is nothing more. But how can they pray who have come to this? Can it ever be useful or helpful to believe a lie? That which has been threatened as the worst of all spiritual evils, would then become the conscious attitude of our "religion," the habitual condition of our worship. This must be as bad science, as it is bad religion. It is in violation of a Law the highest known to Man—the Law which inseparably connects earnest conviction of the truth in what we do or say, with the very fountains of all intellectual and moral strength. No accession of force can come to us from doing anything in which we disbelieve. Such a doctrine will be indeed

> "The little rift within the lute
> That by and by will make the music mute,
> And ever widening slowly silence all." [1]

If there is any helpfulness in Prayer even to the Mind itself, that helpfulness can only be preserved by showing that the belief on which this virtue depends is a rational belief. The very essence of that belief is this—that the Divine Mind is accessible to supplication, and that the Divine Will is capable of being moved

[1] "Idylls of the King—Vivien."—Tennyson.

thereby. No question is, or indeed can be, raised as to the powerful effect exerted by this belief on Man's nature. That effect is recognised as a fact. Its value is admitted; and in order that it may not be lost, the compromise now offered by some philosophers is this—that although the course of external nature is unalterable, yet possibly the phenomena of Mind and character may be changed by the Divine Agency. But will this reasoning bear analysis? Can the distinction it assumes be maintained? Whatever difficulties there may be in reconciling the ideas of Law and of Volition, they are difficulties which apply equally to the Worlds of Matter and of Mind. The Mind is as much subject to Law as the Body is. The Reign of Law is over all; and if its dominion be really incompatible with the agency of Volition, Human or Divine, then the Mind is as inaccessible to that agency as material things. It would indeed be absurd to affirm that all Prayers are equally rational or equally legitimate. Most true it is that "we know not what we should pray for as we ought." Prayer does not require us to believe that anything can be done without the use of means; neither does it require us to believe that anything will be done in violation of the Universal Order. "If it be possible," was the qualification used in the most solemn Prayer ever uttered upon Earth. What are and what are not legitimate objects of supplication, is a question which may well be open. But the question now raised is a wider one than this—even the question whether the very idea of Prayer be not in itself absurd—whether the Reign of Law does not preclude the possibility of Will affecting the successive phenomena either of Matter or of Mind. This is a question lying at the root of our whole conceptions of the Universe, and of all our own powers, both of thinking and of acting. The freedom which is denied to God is not likely to be left to Man. We shall see, accordingly, that precisely the same denials are applied to both.

The conception of Natural Laws—of their place, of their nature, and of their office—which involves us in such questions, and which points to such conclusions, demands surely a very careful examination at our hands.

What, then, is this Reign of Law? What is Law, and in what sense can it be said to reign?

Words, which should be the servants of Thought, are too often its masters; and there are very few words which are used more ambiguously, and therefore more injuriously, than the word "Law." It may indeed be legitimately used in several different senses, because in all cases as applied in Science it is a metaphor, and one which has relation to many different kinds and degrees of likeness in the ideas which are compared. It matters little in which of these senses it is used, provided the distinctions between them are kept clearly in view, and provided we watch against the fallacies which must arise when we pass insensibly from one meaning to another. And here it may be observed, in passing, that the metaphors which are employed in Language are generally founded on analogies instinctively, and often unconsciously, perceived, and which would not be so perceived if they were not both deep and true. In this case the idea which lies at the root of Law in all its applications is evident enough. In its primary signification, a "law" is the authoritative expression of human Will enforced by Power. The instincts of mankind finding utterance in their language, have not failed to see that the phenomena of Nature are only really conceivable to us as in like manner the expressions of a Will enforcing itself with Power. But, as in many other cases, the secondary or derivative senses of the word have supplanted the primary signification; and Law is now habitually used by men who deny the analogy on which that use is founded, and to the truth of which it is an abiding witness. It becomes therefore all the more necessary to define the secondary senses with precision. There are at least Five different senses in which Law is habitually used, and these must be carefully distinguished:—

First, We have Law as applied simply to an observed Order of facts.

Secondly, To that Order as involving the action of some Force or Forces of which nothing more may be known.

Thirdly, As applied to individual Forces the measure of whose operation has been more or less defined or ascertained.

Fourthly, As applied to those combinations of Force which have reference to the fulfilment of Purpose, or the discharge of Function.

Fifthly, As applied to Abstract Conceptions of the mind—not corresponding with any actual phenomena, but deduced therefrom as axioms of thought necessary to our understanding of them. Law, in this sense, is a reduction of the phenomena, not merely to an Order of facts, but to an Order of Thought.

These great leading significations of the word Law all circle round the three great questions which Science asks of Nature, the What, the How, and the Why :—

(1) What are the facts in their established Order?

(2) How—that is, from what physical causes,—does that Order come to be?

(3) Why have these causes been so combined? What relation do they bear to Purpose, to the fulfilment of intention, to the discharge of Function?

It is so important that these different senses of the word Law should be clearly distinguished, that each of them must be more fully considered by itself.

The First and, so to speak, the lowest sense in which Law is applied to natural phenomena is that in which it is used to express simply "an observed Order of facts"—that is to say, facts which under the same conditions always follow each other in the same order. In this sense the laws of Nature are simply those facts of Nature which recur according to a rule. It is not necessary to the legitimate application of Law in this sense, that the cause of any observed Order of facts should be at all known, or even guessed at. The Force or Forces to which that Order is due may be hid in total darkness. It is sufficient that the Order or sequence of phenomena be uniform and constant. The neatest and simplest illustration of this, as well as of the other senses in which Law is used, is to be found in the exact sciences, and especially in the history of Astronomy. It is nearly 250 years since Kepler discovered, in respect to the distances, velocities, and orbits of the Planets, three facts, or rather three series of facts, which, during many years[1] of intense application to physical inquiry, remained the highest truths known to Man on the pheno-

[1] The "Third Law" of Kepler was made known to the world in 1619. Newton's "Principia" appeared in 1687.

mena of the Solar System. They were known as the Three Laws of Kepler. It is not necessary to describe in detail here what these laws were. Suffice it to say, that the most remarkable among them were facts of constant numerical relation between the distances of the different Planets from the Sun, and the length of their periodic times; and again, between the velocity of their motion and the space enclosed within certain corresponding sections of their orbit. These Laws were simply and purely an "Order of facts" established by observation, and not connected with any known cause. The Force of which that Order is a necessary result had not then been ascertained. A very large proportion of the laws of every science are laws of this kind and in this sense. For example, in Chemistry the behaviour of different substances towards each other, in respect to combination and affinity, is reduced to system under laws of this kind, and of this kind only. Because, although there is a probability that Electric or Galvanic Force is the cause, or one of the causes, of the series of facts exhibited in chemical phenomena, this is as yet no better than a probability, and the laws of Chemistry stand no higher than facts which by observation and experiment are found to follow certain rules.

But the ascertainment of a law in this First and lower sense leads immediately and instinctively to the search after Law in another sense which is higher. An observed Order of facts, to be entitled to the rank of a Law, must be an Order so constant and uniform as to indicate necessity, and necessity can only arise out of the action of some compelling Force. Law, therefore, comes to indicate not merely an observed Order of facts, but that Order as involving the action of some Force or Forces, of which nothing more may be known than these visible effects. Every observed Order in physical phenomena suggests irresistibly to the mind the operation of some physical cause. We say of an observed Order of facts that it must be due to some "law," meaning simply that all Order involves the idea of some arranging cause, the working of some Force or Forces (whether they be such as we can further trace and define or not) of which that Order is the index and the result. This is the Second of the five senses specified above.

And so we pass on, by an easy and natural transition, to the Third sense in which the word Law is used. This is the most exact and definite of all. The mere general idea that some Force is at the bottom of all phenomena, which are invariably consecutive, is a very different thing from knowing what that Force is in respect to the rule or measure of its operation. Of Law in this sense the one great example, before and above all others, is the Law of Gravitation, for this is a Law in the sense not merely of a rule, but of a cause—that is, of a Force accurately defined and ascertained according to the measure of its operation, from which Force other phenomena arise by way of necessary consequence. Force is the root-idea of Law in its scientific sense. And so the Law of Gravitation is not merely the "observed order" in which the heavenly bodies move; neither is it only the abstract idea of some Force to which such movements must be due, but it is that Force the exact measure of whose operation was numerically ascertained or defined by Newton — the Force which compels those movements and (in a sense) explains them. Now the difference between Law in the narrower and Law in the larger sense cannot be better illustrated than in the difference between the Three special Laws discovered by Kepler, and the One universal Law discovered by Newton. The Three Laws of Kepler were, as we have seen, simply and purely an observed Order of facts. They stood by themselves —disconnected,—their cause unknown. The higher Law, discovered by Newton, revealed their connexion and their cause. The "observed Order" which Kepler had discovered, was simply a necessary consequence of the Force of Gravitation. In the light of this great Law the "Three Laws of Kepler" have been merged and lost,

When the operations of any material Force can be reduced to rules so definite as those which have been discovered in respect to the Force of Gravitation, and when these rules are capable of mathematical expression and of mathematical proof, they are, so far as they go, in the nature of pure truth. Mr. Lewes, in his very curious and interesting work on the "Philosophy of Aristotle," has maintained that the knowledge of Measure—or what he calls the "verifiable element" in our knowledge—is the element which determines whether any theory belongs to Science, strictly so called, or to Metaphysics; and that any theory may be transferred from Metaphysics to Science, or from Science to Metaphysics, simply by the addition or withdrawal of its "verifiable element." In illustration of this, he says that if we withdraw, from the Law of Universal Attraction, the formula, "inversely as the square of the distance, and directly as the mass," it becomes pure Metaphysics. If this means that, apart from ascertained numerical relations, our conception of Law, or our knowledge of natural phenomena, loses all reality and distinctness, I do not agree in the position. The idea of natural Forces is quite separate from any ascertained measure of their energy. The knowledge, for example, that all the particles of matter exert an attractive force upon each other, is, so far as it goes, true physical knowledge, even though we did not know the further truth, that this force acts according to the numerical rule ascertained by Newton. To banish from physical Science, properly so called, and to relegate to Metaphysics, all knowledge which cannot be reduced to numerical expression, is a dangerous abuse of language.

Force, ascertained according to some measure of its operation—this is indeed one of the definitions, but only one, of a scientific Law. The discovery of laws in this sense is the great quest of Science, and the finding of them is one of her great rewards. Such laws yield to the human mind a peculiar delight, from the satisfaction they afford to those special faculties whose function it is to recognise the beauty of numerical relations. This satisfaction is so great, and in its own measure is so complete, that the mind reposes on an ascertained law of this kind as on an ultimate truth. And ultimate it is as regards the particular faculties which are concerned in this kind of search. When we have observed our facts, and when we have summed up our figures, when we have recognised the constant numbers,—then our eyes, our ears, and our calculating faculties have done their work. But other faculties are called into simultaneous operation, and these have other work to do. For let it be observed that laws, in the first three senses we have now examined, cannot be said to explain anything except the Order of subordinate phenomena. They set forth that order as due to Force. They do nothing more. Least of all do laws, in any of these three senses, explain themselves. They suggest a thousand questions much more curious than the questions which they solve. The very beauty and simplicity of some laws is their deepest mystery. What can their source be? How is their uniformity maintained? Every law implies a Force, and all that we ever know is some numerical rule or measure according to which some unknown Forces operate. But whence come those measures—those exact relations to number, which never vary? Or, if there are variations, how comes it that these are always found to follow some other rules as exact and as invariable as the first?

And as there can be no better example of what Law is, so also there can be no better example of what it is not—than the Law of Gravitation. The discovery of it was probably the highest exercise of pure intellect through which the human mind has found its way. It is the most universal physical law which is known to us, for it prevails, apparently, through all Space. Yet of the Force of Gravitation all we know is, that it is a force of attraction operating between all the particles of matter in the exact measure which was ascertained by

Newton,—that is—"directly as the mass, and inversely as the square of the distance." This is the Law. But it affords no sort of explanation of itself. What is the cause of this Force—what is its source—what are the media of its operation—how is the exact uniformity of its proportions maintained?—these are questions which it is impossible not to ask, but which it is quite as impossible to answer. Sir John Herschel, in speaking of this Force, has indicated in a passing sentence a few questions out of the many which arise:—"No matter," he says, "from what ultimate causes the power called gravitation originates—be it a virtue lodged in the sun as its receptacle, or be it pressure from without, or the resultant of many pressures, or solicitations of unknown kinds, magnetic or electric, ethers or impulses,"[1] &c. &c. How little we have ascertained in this Law, after all! Yet there is an immense and an instinctive pleasure in the contemplation of it. To analyse this pleasure is as difficult as to analyse the pleasure which the eye takes in beauty of form, or the pleasure which the ear takes in the harmonies of sound. And this pleasure is inexhaustible, for these laws of number and proportion pervade all Nature, and the intellectual organs which have been fitted to the knowledge of them have eyes

[1] Herschel's "Outlines of Astronomy," fifth edition, p. 323.

which are never satisfied with seeing, and ears which are never full of hearing. The agitation which overpowered Sir Isaac Newton as the Law of Gravitation was rising to his view in the light of rigorous demonstration, was the homage rendered by the great faculties of his nature to a harmony which was as new as it was immense and wonderful. The same pleasure in its own degree is felt by every man of science who, in any branch of physical inquiry, traces and detects any lesser law. And it is perfectly true that such laws are being detected everywhere. Forces which are in their essence and their source utterly mysterious, are always being found to operate under rules which have strict reference to measures of number,—to relations of Space and Time. The Forces which determine chemical combination all work under rules as sharp and definite as the Force of Gravitation. So do the Forces which operate in Light, and Heat, and Sound. So do those which exert their energies in Magnetism and Electricity. All the operations of Nature—the smallest and the greatest—are performed under similar measures and restraints. Not even a drop of water can be formed except under rules which determine its weight, its volume, and its shape, with exact reference to the density of the fluid, to the structure of the surface on which it may be formed, and to the pressure of the surrounding atmosphere. Then that pressure is itself exercised under rigorous rules again. Not one of the countless varieties of form which prevail in clouds, and which give to the face of heaven such infinite expression, not one of them but is ruled by Law,—woven, or braided, or torn, or scattered, or gathered up again and folded,—by Forces which are free only "within the bounds of Law."

And equally in those subjects of inquiry in which rules of number and of proportion are not applicable, rules are discernible which belong to another class, but which are as certain and as prevailing. All events, however casual or disconnected they may at first appear to be, are found in the course of time to arrange themselves in some certain Order, the index and exponent of Forces, of which we know nothing except their existence as evidenced in these effects. It is indeed wonderful to find that in such a matter, for example, as the development of our Human Speech, the unconscious changes which arise from time to time among the rudest utterances of the rudest tribes and races of Mankind, are all found to follow rules of progress as regular as those which preside over any of the material growths of Nature. Yet so it is; and it is upon this fact alone that the science of Language rests—a science in which all the facts are not yet observed, and many of those which have been observed are not yet reduced to order; but in which enough has been ascertained to show that languages grow, and change from generation to generation, according to rules of which the men who speak them are wholly unconscious. It is the same with all other things. And as it is now, so apparently has it been in all past time of which we have any record. Even the work of Creation has been and is being carried on under rules of adherence to Typical Forms, and under limits of variation from them, which can be dimly seen and traced, although they cannot be defined or understood. The universal prevalence of laws of this kind cannot therefore be denied. The discovery of them is one of the first results of all physical inquiry. In this sense it is true that we, and the world around us, are under the Reign of Law.

It is true, but only a bit and fragment of the truth. For there is another fact quite as prominent as the universal presence and prevalence of laws—and that is, the number of them which are concerned in each single operation in Nature. No one Law—that is to say, no one Force—determines anything that we see happening or done around us. It is always the result of different and opposing Forces nicely balanced against each other. The least disturbance of the proportion in which any one of them is allowed to tell, produces a total change in the effect. The more we know of Nature, the more intricate do such combinations appear to be. They can be traced very near to the fountains of Life itself, even close up to the confines of the last secret of all—how the Will acts upon its organs in the Body. Recent investigations in Physiology seem to favour the hypothesis that our muscles are the seat of two opposing Forces, each so adjusted as to counteract the other; and that this antagonism is itself so arranged as to enable us by acting on one of these Forces, to regulate the action of the other. One Force—an elastic or contractile Force—is supposed to be inherent in the muscular fibre: another Force—that of Animal Electricity in statical condition—holds the contractile Force in check; and the relaxed, or rather the restful, condition of the muscle when not in use, is due to the balance so maintained. When, through the motor nerves the Will orders the muscles into action, that order is enforced by a discharge of the Electrical Force, and upon this discharge the contractile Force is set free to act, and does accordingly produce the contraction which is desired.[1]

Such is, at least, one suggestion as to the means employed to place human action under the control of

1 This theory of muscular and nervous action is set forth with much ingenuity and force of illustration in "Lectures on Epilepsy," &c., by Charles Bland Radcliffe, M.D.

human Will, in that material frame which is so wonderfully and fearfully made. And whether this hypothesis be accurate or not, it is certain that some such adjustment of Force to Mechanism is involved in every bodily movement which is subject to the Will. Even in this high region, therefore, we see that the existence of individual laws is not the end of our physical knowledge. What we always reach at last in the course of every physical inquiry, is the recognition, not of individual laws, but of some definite relation to each other, in which different laws are placed, so as to bring about a particular result. But this is, in other words, the principle of Adjustment, and adjustment has no meaning except as the instrument and the result of Purpose. Force so combined with Force as to produce certain definite and orderly results,—this is the ultimate fact of all discovery.

And so we come upon another sense—the Fourth sense, in which Law is habitually used in Science, and this is perhaps the commonest and most important of all. It is used to designate not merely an observed Order of facts—not merely the bare abstract idea of Force—not merely individual Forces according to ascertained measures of operation—but a number of Forces in the condition of mutual adjustment, that is to say, as combined with each other, and fitted to each other for the attainment of special ends. The whole science of Animal Mechanics, for example, deals with Law in this sense—with natural Forces as related to Purpose and subservient to the discharge of Function. And this is the highest sense of all—Law in this sense being more perfectly intelligible to us than in any other; because, although we know nothing of the real nature of Force, even of that Force which is resident in ourselves, we do know for what ends we exert it, and the principle that governs our devices for its use. That principle is, Combination for the accomplishment of Purpose.

Accordingly it is, when natural phenomena can be reduced to Law, in this last sense, that we reach something which alone is really in the nature of an explanation. For what do we mean by an explanation? It is an unfolding or a "making plain." But as the human mind has many faculties, so each of these seeks a satisfaction of its own. That which is made plain to one faculty is not necessarily made plain to another. That which is a complete answer to the question What, or to the question How, is no answer at all to the question Why. There are some philosophers who tell us that this last is a question which had better never be asked, because it is one to which Nature gives no reply. If this be so, it is strange that Nature should have given us the faculties which impel us to ask this question—ay, and to ask it more eagerly than any other. It is, indeed, true that there is a point beyond which we need not ask it, because the answer is inaccessible. But this is equally true of the questions What, and How. We cannot reach Final Causes any more than Final Purposes. For every cause which we can detect, there is another cause which lies behind; and for every purpose which we can see, there are other purposes which lie beyond.

And so it is true that all things in Nature may either be regarded as means or as ends—for they are always both—only that Final Ends we can never see. For, as Bishop Butler truly says in his "Analogy,"[1] "We know what we ourselves aim at as final ends, and what courses we take merely as means conducing to these ends. But we are greatly ignorant how far things are considered by the Author of Nature under the simple notion of means and ends,—so as that it may be said this is merely an end, and that merely means, in His regard. And whether there be not some peculiar absurdity in our very manner of conception concerning this matter, somewhat contradictory, arising from an extremely imperfect view of things, it is impossible to say." This is indeed a wise caution, and one which has been much needed to check the abuse of that method of reasoning which has been

1 Butler's "Analogy," chap. iv.

called the doctrine of Final Causes. When Man makes an implement, he knows the purpose for which he makes it—he knows the function assigned to it in his own intention. But as in making it there are a thousand chips and fragments of material which he casts aside, so in its final use it often produces consequences and results which he did not contemplate or foresee. But in Nature all this is different. Nature has no chips or fragments which she does not put to use; and as on the way to her apparent ends there are no incidents which she did not foresee, so beyond those ends there are no ulterior results which do not open out into new firmaments of Design. Of nothing, therefore, can we say with even the probability of truth that we see its Final Cause; that is to say, its ultimate purpose. All that we can ever see are the facts of Adjustment and of Function, and these constitute not Final, but Immediate Purpose. But a purpose is not less a purpose, because other purposes may lie beyond it. And not only can we detect Purpose in natural phenomena, but, as we have already seen, it is very often the only thing about them which is intelligible to us. The How is very often incomprehensible, where the Why is apparent at a glance. And be this observed, that when Purpose is perceived, it is a "making plain" to a higher faculty of the mind than the mere sense of Order. It is a making plain to Reason. It is the reduction of phenomena to that Order of Thought which is the basis of all other Order in the works of Man, and which, he instinctively concludes, is the basis also of all Order in the works of Nature.

And here it is important to observe, that although this general conclusion, like all other general conclusions, belongs to the category of mental inferences, and not to the category of physical facts, yet each particular instance of Purpose on which the general inference is founded, is not an inference merely, but a fact. The function of an organ, for example, is a matter of purely physical investigation. But the function of an organ is not merely that which it does, but it is that which some special construction enables it to do. It is, not merely its work, but it is the work assigned to it as an Apparatus, and as fitted to other organs having other functions related to its own. The nature of that Apparatus, as being in itself an adjustment for a particular purpose, is not an inference from the facts, but it is part of the facts themselves. The very idea of Function is inseparable from the idea of Purpose. The Function of an organ is its Purpose ; and the relation of its parts, and of the whole to that Purpose, is as much and as definitely a scientific fact as the relation of any other phenomenon to Space, or Time, or Number.

This distinction between Purpose as a general inference and Purpose as a particular fact, has not been sufficiently observed. The just condemnation pronounced by Bacon on the pursuit of Final Causes as distorting the true Method of Physical Investigation, has been applied without discrimination to two very different conceptions. Even Philosophers who believe in the Supremacy of Purpose in Nature have been willing to banish this conception from the Domain of Science, and to classify it as belonging altogether to Metaphysics or Theology. Thus in the very able Harveian Oration for 1865 by Dr. H. W. Acland, he says, —"Whether there be any Purpose, is the object of Theological and Metaphysical, but not of Physical inquiry."[1] And again, "The evidence of intention is metaphysical, and depends on probabilities. It is not positive. It is inferential from many considerations."[2] I venture to dissent from these conclusions. They involve, I think, a confounding of two separate questions. The nature and character of the intending Mind—this is indeed a question of Theology; but not the existence of intention. Neither in any restrictive sense of the word can it be called Metaphysical. Even as a general doctrine, the doctrine of Contrivance and

[1] P. 61. [2] P. 63.

Adjustment is not so metaphysical as the Doctrine of Homologies; and when we come to particular cases there can be no question whatever that the relation of a given Structure to its Purpose and Function comes more unequivocally under the class of physical facts than the relation of that same Structure to some corresponding part in another animal. It is less ideal, for example,—less theoretical—less metaphysical—to assert of the little hooked claw which is attached to the (apparent) elbow of a Bat's wing, that it was placed there to enable the Bat to climb and crawl, than to affirm of that same claw that it is the "homologue" of the human thumb. Yet who can deny that this doctrine of Homologies has been established as a strictly scientific truth? There is a sense, of course, in which all Knowledge and all Science belongs to Metaphysics. Mere classification, which is the basis of all Science, what is it but the marshalling of physical facts in an Ideal Order—an arrangement of them according to the relation which they bear to the laws of Thought? But this does not constitute as a branch of Metaphysics, the division of animals into Genera, and Families, and Orders. And what relation can physical facts ever have to Thought so directly cognisable or so susceptible of Demonstration as the relation of an animal organ to its purpose and function in the animal economy?

Whether Purpose be the basis of all natural Order or not is a separate question. It is at least one of the facts of that Order. Combination for the accomplishment of Purpose therefore in particular cases, such as the relation between the structure of an Organ and its function, is not merely a safe conclusion of Philosophy, but an ascertained fact of Science.[1]

This question has acquired additional importance since the revival in our own day, and with new resources, of that old philosophy which assumes to banish from the domain of Knowledge no small part of the richest and surest acquisitions of Reason. That Philosophy must be tested by a rigid analysis of thought and language. This is the weapon with which the assault is made, and it is by the same weapon better handled that it can alone be met. An arbitrary limitation of the word "knowledge," to a particular kind of knowledge, can only be tolerated on condition that the arbitrary nature of the limitation be constantly kept in view. In like manner the word "verification" may be confined to a particular kind of proof applicable only to a particular class of truths. So again, in regard to "Metaphysics," it may be considered with reference to its subject-matter as denoting a particular branch of

[1] See Note B.

inquiry—such as Psychology—or as a method of investigation which may be applied equally to all subjects which furnish the mind with the materials of thought. But we must watch against the substitution of one of these meanings for another; and against the jugglery by which men first use Metaphysical Analysis to pull down conceptions which they dislike, and then denounce Metaphysics as incapable of establishing any conclusions on which we can rely. The fact to which I have previously referred,[1] is a fact of immense significance, that one of the most able supporters of the Positive Philosophy in England relegates to Metaphysics the great scientific fact of Physical Attraction, when it is considered apart from its numerical relations. But if this be considered Metaphysics, then let it be remembered that many of the most certain truths we know belong to the same category. From a similar point of view, it might be argued, and it has actually been argued, that Number and all numerical relations are purely abstract conceptions of the mind, having no other reality than as there conceived.[2] The same reasoning may be applied to all our most fundamental conceptions—without which Science could not even begin her work. The existence of Force under any form, of which the existence of Matter

[1] P. 70. [2] See Note C.

is only a special case, may be regarded as a purely metaphysical conception. It is surely a comfort to find that, if all ideas of Plan and of Design in the Adjustments of Organic Life are to be condemned as Metaphysical, they stand at least in goodly company among the necessities of Thought. Mr. Lewes, indeed, himself confesses that "Science finds it indispensable to co-ordinate all the facts in a general concept, such as a Plan."[1] But he pronounces it one of the "Infirmities of Thought" to "realize the concept." But no accurate thinker ever "realized" such an idea as a "Plan"—that is to say, no one ever conceived it as existing by itself, separate from an intending Mind. Mr. Lewes complains that "Matter and Force are mysterious enough" without a "new mystery of Architectural Plan, shaping Matter and directing Force."[2] But, substituting here "Mind" for Plan, it may surely be argued that, if Science finds it "indispensable" to co-ordinate all the facts in some such general concept, this is of itself a proof that the element so introduced does not add to the mystery, but helps to remove it. Even if it be an "artifice of thought," it can only be resorted to as rendering the facts not less but more conceivable. And this it plainly does by appealing to

[1] "History of Philosophy," Prologue, p. lxxxvi. [2] Ibid.

an agency having known power in the production of analogous phenomena. The instinctive wisdom which lies in this "infirmity" of the mind becomes more apparent when we turn to the efforts of an acute intellect to cast such infirmities away. The most abstract metaphysical conceptions are substituted for those which are denounced: the only difference being that, whilst the old conceptions are intelligible as connecting the Phenomena by a link of thought which the mind can feel and follow, the new conceptions are unintelligible because they try to describe facts without any reference to the ideas they involve. No new light—nothing but denser darkness—is cast on the phenomena of Organic Life by calling "Life the connexus of the organic activities."[1] Yet meaningless words are heaped on each other in the desperate effort to dispense with those conceptions which can alone render the order of Nature intelligible to us. Thus we are told again, that "The Organism is the synthesis of diverse parts, and Life is the synthesis of their properties;"[2]—and again, that "Vitality is the abstract designation of certain special properties manifested by Matter under certain special conditions."[3] Surely there is more light in the old reading:—"Finding," says Mr. Lewes, "in an organism

[1] "History of Philosophy," p. lxxx. [2] Ibid. p. lxxxiii.
[3] Ibid. p. lxxxiv.

a certain adjustment of parts, which may be reduced to a plan, we are *easily led to conceive* that this plan was made before the parts, and that the adjustment was determined by the plan." No doubt! This is the easiest conception, and it is the easiest because it is most conformable to the laws of Thought; and that which is the most conformable to the laws of Thought is that which makes the nearest approach to absolute Truth attainable by the Mind.

The universal prevalence of this idea of Purpose in Nature is indicated by the irresistible tendency which we observe in the language of Science to personify the Forces, and the combinations of Force by which all natural phenomena are produced. It is a great injustice to scientific men—too often committed—to suspect them of unwillingness to accept the idea of a Personal Creator merely because they try to keep separate the language of Science from the language of Theology.[1] But it is

[1] A remarkable instance of this injustice has been lately brought to light. Professor Huxley, in an article in the *Fortnightly Review*, had used one of those vague phrases, so common with scientific men, about the "unknown and the unknowable" being the goal of all scientific thought, which not unnaturally suggest the notion that all idea of a God is unattainable. A writer in the *Spectator* accordingly dealt with Professor Huxley as avowing Atheism, and was rebuked by the Professor in a letter published in the *Spectator* of Feb. 10, 1866. Professor Huxley says: "I do not know that I care very much about popular odium, so that there is no great merit

curious to observe how this endeavour constantly breaks down—how impossible it is in describing physical phenomena to avoid the phraseology which identifies them with the phenomena of Mind, and is moulded on our own conscious Personality and Will. It is impossible to avoid this language simply because no other language conveys the impression which innumerable structures leave upon the mind. Take, for example, the word "contrivance." How could Science do without it? How could the great subject of Animal Mechanics be dealt with scientifically without continual reference to Law as that by which, and through which, special organs are formed for the doing of special work? What is the very definition of a machine? Machines do not increase Force, they only adjust it. The very idea and essence of a machine is that it is a contrivance for the distribution of Force with a view

in saying that if I really saw fit to deny the existence of a God, I should certainly do so, for the sake of my own intellectual freedom, and be the honest Atheist you are pleased to say I am. As it happens, however, I cannot take this position with honesty, inasmuch as it is, and always has been, a favourite tenet of mine, that Atheism is as absurd, logically speaking, as Polytheism." On the subject of miracles, in the same letter, Professor Huxley says, that "denying the possibility of miracles seems to me quite as unjustifiable as speculative Atheism." The question of miracles seems now to be admitted on all hands to be simply a question of evidence.

to its bearing on special purposes. A man's arm is a machine in which the law of leverage is supplied to the vital force for the purposes of prehension. We shall see presently that a bird's wing is a machine in which the same law is applied, under the most complicated conditions, for the purpose of flight. Anatomy supplies an infinite number of similar examples. It is impossible to describe or explain the facts we meet with in this or in any other branch of Science without investing the "laws" of Nature with something of that Personality which they do actually reflect, or without conceiving of them as partaking of those attributes of Mind which we everywhere recognise in their working and results.

We may, again, take the Forces which determine the Planetary motions as the grandest and the simplest illustrations of this truth of Science. Gravitation, as already said, is a Force which prevails apparently through all Space. But it does not prevail alone. It is a Force whose function it is to balance other Forces, of which we know nothing, except this,—that these, again, are needed to balance the Force of Gravitation. Each Force, if left to itself, would be destructive of the Universe. Were it not for the Force of Gravitation, the centrifugal Forces which impel the Planets would fling them off into Space. Were it not for these centrifugal Forces, the Force of Gravitation would dash them against the Sun. The orbits, therefore, of the Planets, with all that depends upon them, are determined by the nice and perfect balance which is maintained between these two Forces; and the ultimate fact of astronomical science is not the Law of Gravitation, but the Adjustment between this law and others which are less known, so as to produce and maintain the existing Solar System.

This is one example of the principle of Adjustment; but no one example, however grand the scale may be on which it is exhibited, can give any idea of the extent to which the principle of Adjustment is required, and is adopted in the works of Nature. The revolution of the seasons, for example—seed-time and harvest—depend on the Law of Gravitation in this sense, that if that law were disturbed, or if it were inconstant, they would be disturbed and inconstant also. But the seasons equally depend on a multitude of other laws,—laws of heat, laws of light, laws relating to fluids, and to solids, and to gases, and to magnetic attractions and repulsions, each one of which laws is invariable in itself, but each of which would produce utter confusion if it were allowed to operate alone, or if it were not balanced against others in the right proportion. It is very difficult to form any adequate idea of the vast number of laws which are concerned in producing the most ordinary operations of Nature. Looking only at the combinations with which Astronomy is concerned, the adjustments are almost infinite. Each minutest circumstance in the position, or size, or shape of the Earth, the direction of its axis, the velocity of its motion and of its rotation, has its own definite effect, and the slightest change in any one of these relations would wholly alter the world we live in. And then it is to be remembered that the seasons, as they are now fitted to us, and as we are fitted to them, do not depend only on the facts or the laws which Astronomy reveals. They depend quite as much on other sets of facts, and other sets of laws, revealed by other sciences, —such, for example, as Chemistry, Electricity, and Geology. The motion of the Earth might be exactly what it is, every fact in respect to our Planetary position might remain unchanged, yet the seasons would return in vain if our own atmosphere were altered in any one of the elements of its composition, or if any one of the laws regulating the action were other than it is. Under a thinner air even the torrid zone might be wrapped in eternal snow. Under a denser air, and one with different refracting powers, the Earth and all that is therein might be burnt up. And so it is through the whole of Nature: laws everywhere—laws in themselves invariable, but so worked as to produce effects of inexhaustible variety by being pitched against each other, and made to hold each other in restraint.

I have already referred to Chemistry as a science full of illustrations of Law in the First and simplest sense—

that is, of facts in observed orders of recurrence. But Chemistry is a science not less rich in illustration of Law in the Fourth sense—that is, of Forces in mutual adjustment. Indeed, in Chemistry, this system of adjustment among the different properties of matter is especially intricate and observable. Some of the laws which regulate Chemical Combination were discovered in our own time, and are amongst the most wonderful and the most beautiful which have been revealed by any science. They are laws of great exactness, having invariable relations to number and proportion. Each elementary substance has its own combining proportions with other elements, so that, except in these proportions, no chemical union can take place at all. And when chemical union does take place, the compounds which result have different and even opposite powers, according to the different proportions employed. Then, the relations in which those inorganic compounds stand to the chemistry of Life, constitute another vast series in which the principle of adjustment has applications, infinite in number, and as infinite in beauty. How delicate these relations are, and how tremendous are the issues depending on their management, may be conceived from this single fact,—that the same elements combined in one proportion are sometimes a nutritious food or a grateful stimulant, soothing and sustaining the powers of life; whilst, combined in another proportion, they may be a deadly poison, paralysing the heart and carrying agony along every nerve and fibre of the animal frame. This is no mere theoretical possibility. It is actually the relation, for example, in which two well-known substances stand to each other—Tea and Strychnia. The active principles of these two substances, "Theine" and "Strychnine," are identical so far as their elements are concerned, and differ from each other only in the proportions in which they are combined. Such is the power of numbers in the Laboratory of Nature! What havoc in this world, so full of Life, would be made by blind chance gambling with such powers as these! What confusion, unless they were governed by laws whose certainty makes them capable of fine adjustment, and therefore subject to accurate control! How fine these adjustments are, and how absolute is that control, is indicated in another fact—and that is the few elements out of which all things are made. The number of substances deemed elementary has varied with the advance of Science; but as compared with the variety of their products, that number may be considered as infinitesimally small; whilst the progress of analysis, with glimpses of laws as yet unknown, renders it almost certain that this number will be found to be smaller still. Yet out of that small number of elementary substances, having fixed rules, too, limiting their combination, all the infinite varieties of organic and inorganic matter are built up by means of nice adjustment. As all the faculties of a powerful mind can utter their voice in language whose elements are reducible to twenty-four letters, so all the forms of Nature, with all the ideas they express, are worked out from a few simple elements having a few simple properties.

Simple! can we call them so? Yes, simple by comparison with the exceeding complication of the uses they are made to serve: simple also, in this sense, that they follow some simple rule of numbers. But in themselves these laws, these forces are incomprehensible. That which is most remarkable about them is their unchangeableness. The whole mind and imagination of scientific men is often so impressed with this character of material laws, that no room is left for the perception of other aspects of their nature and of their work. We hear of rigid and universal sequence—necessary—invariable;—of unbroken chains of cause and effect, no link of which can, in the nature of things, be ever broken. And this idea grows upon the mind, until in some confused manner it is held as casting out the idea of Purpose in creation, and inconsistent with the element of Will. If it be so, the difficulty cannot be evaded by denying the uniformity, any more than the universality, of Law. It is perfectly true that every law is, in its own nature, invariable, producing always precisely and necessarily the same effects,—that is, provided it is worked under the same conditions. But then, if the conditions are not the same, the invariableness of effect gives place to capacities of change which are almost infinite. It is by altering the conditions under which any given law is brought to bear, and by bringing other laws to operate upon the same subject, that our own Wills exercise a large and increasing power over the material world. And be it observed—to this end the uniformity of laws is no impediment, but, on the contrary, it is an indispensable condition. Laws are in themselves—if not unchangeable—at least unchanging, and if they were not unchanging, they could not be used as the instruments of Will. If they were less rigorous they would be less certain, and the least uncertainty would render them incapable of any service. No adjustment, however nice, could secure its purpose if the implements employed were of uncertain temper.

The notion therefore that the uniformity or invariableness of the Laws of Nature cannot be reconciled with their subordination to the exercise of Will, is a notion contrary to our own experience. It is a confusion of thought arising very much out of the ambiguity of language. For let it be observed that, of all the senses in which the word Law is used, there is only one in which it is true that laws are immutable or invariable; and that is the sense in which Law is used to designate an individual Force. Gravitation, for example, is immutable in this respect—that (so far as we know) it never operates according to any other measure than "directly as the mass, and inversely as the square of the distance." But in all the other senses in which the word Law is used, laws are not immutable; but, on the contrary, they are the great instruments, the unceasing agencies, of change. When, therefore, scientific men speak, as they often do, of all phenomena being governed by invariable laws, they use language which is ambiguous, and in most cases they use it in a sense which covers an erroneous idea of the facts. There are no phenomena visible to Man of which it is true to say that they are governed by any invariable Force. That which does govern them is always some variable combinations of invariable forces. But this makes all the difference in reasoning on the relation of Will to Law,—this is the one essential distinction to be admitted and observed. There is no observed Order of facts which is not due to a combination of Forces; and there is no combination of Forces which is invariable—none which are not capable of change in infinite degrees. In these senses—and these are the common senses in which Law is used to express the phenomena of Nature—Law is not rigid, it is not immutable, it is not invariable, but it is, on the contrary, pliable, subtle, various. In the only sense in which laws are immutable, this immutability is the very characteristic which makes them subject to guidance through endless cycles of design. We know this in our own case. It is the very certainty and invariableness of the laws of Nature which alone enables us to use them, and to yoke them to our service.

Now, the laws of Nature appear to be employed in the system of Nature in a manner precisely analogous to that in which we ourselves employ them. The difficulties and obstructions which are presented by one law in the way of accomplishing a given purpose, are met and overcome exactly on the same principle on which they are met and overcome by Man—viz., by knowledge of other laws, and by resource in applying them,—that is, by ingenuity in mechanical contrivance. It cannot be too much insisted on, that this is a conclusion of pure Science. The relation which an organic structure bears to its purpose in Nature can be recognised as certainly as the same relation between a machine and its purpose in human art. It is absurd to maintain, for example, that the purpose of the cellular arrangement of material in combining lightness with strength, is a purpose legitimately cognisable by Science in the Menai Bridge, but is not as legitimately cognisable when it is seen in Nature, actually serving the same use. The little Barnacles which crust the rocks at low tide, and which to live there at all must be able to resist the surf, have the building of their shells constructed strictly with reference to this necessity. It is a structure all hollowed and chambered on the plan which engineers have so lately discovered as an arrangement of material by which the power of resisting strain or pressure is multiplied in an extraordinary degree. That shell is as pure a bit of mechanics as the bridge, both being structures in which the same arrangement is adapted to the same end.

> "Small, but a work divine;
> Frail, but of force to withstand,
> Year upon year, the shock
> Of cataract seas that snap
> The three-decker's oaken spine."[1]

This is but one instance out of a number which no man can count. So far as we know, no Law—that is, no elementary Force—of Nature is liable to change. But every Law of Nature is liable to counteraction; and the rule is, that laws are habitually made to counteract each other in precisely the manner and degree which some definite result requires.

Nor is it less remarkable that the converse of this is true: no Purpose is ever attained in Nature, except by the enlistment of Laws as the means and instruments of

[1] "Maud."

attainment. When an extraordinary result is aimed at, it often happens that some common law is yoked to extraordinary conditions, and its action is intensified by some special machinery. For example, the Forces of Electricity are in action, probably, in all living Organisms, but certainly in the muscular and nervous system of the higher animals. In a very few (so far as yet known, in only a very few animals among the millions which exist, and these all belonging to the Class of Fishes), the electrical action has been so stored and concentrated as to render it serviceable as a weapon of offence. Creatures which grovel at the bottom of the sea, or in the slime of rivers, have been gifted with the astonishing faculty of wielding at their will the most subtle of all the powers of Nature. They have the faculty of "shooting out lightning" against their enemies or their prey. But this gift has not been given without an exact fulfilment of all the laws which govern Electricity, and which especially govern its concentration and destructive force. The Electric Ray, or Torpedo, has been provided with a Battery closely resembling, but greatly exceeding in the beauty and compactness of its structure, the Batteries whereby Man has now learned to make the laws of Electricity subservient to his will. There are no less than 940 hexagonal columns in this Battery like those of a bees' comb, and each of these is subdivided by a series of horizontal plates, which appear to be analogous to the plates of the Voltaic Pile. The whole is supplied with an enormous amount of nervous matter, four great branches of which are as large as the animal's spinal cord, and these spread out in a multitude of thread-like filaments round the prismatic columns, and finally pass into all the cells.[1] This, again, seems to suggest an analogy with the arrangement by which an electric current, passing through a coil and round a magnet, is used to intensify the magnetic force. A complete knowledge of all the mysteries which have been gradually unfolded from the days of Galvani to those of Faraday, and of many others which are still inscrutable to us, is exhibited in this structure. The laws which are appealed to in the accomplishment of this purpose are many and very complicated; because the conditions to be satisfied refer not merely to the generation of Electric force in the animal to which it is given, but to its effect on the nervous system of the animals against which it is to be employed, and to the conducting medium in which both are moving.

When we contemplate such a structure as this, the idea is borne in with force upon the mind, that the need of conforming to definite conditions seems as absolute a

[1] Owen's "Lectures on Comp. Anat." vol. ii. (Fishes).

necessity in making an Electric Fish as in making an Electric Telegraph. But the fact of these conditions existing, and requiring to be satisfied,—or, in other words, the fact of so many natural laws demanding a first obedience,—is not the ultimate fact, it is not even the main fact, which Science apprehends in such phenomena as these. On the contrary, that which is most observable and most certain, is the manner in which these conditions are met, complied with, and, by being complied with, are overcome. But this is, in other words, the subordination of many laws to a difficult and curious Purpose,—a subordination which is effected through the instrumentality of a purely mechanical contrivance.

It is no objection to this universal truth, that the machines thus employed in Nature are themselves constructed through the agency of Law. They grow—or, in modern phraseology, they are developed. But this makes no difference in the case—or, rather, it only carries us farther back to other and yet other illustrations of the same truth. This is precisely one of those cases already referred to, in which Causes are unknown, whilst Purposes are clear and certain. The Battery of an Electric Fish is both a means and an end. As respects the electric laws which it puts in motion—that is, as respects the Force which it concentrates—it must be regarded as a means. As respects the organic laws by which it is itself developed, it is an end.

What we do know in this case is *why* the apparatus was made; that is to say, what we do know is the Purpose. What we do not know, and have no idea of, is *how* it was made; that is to say, what we do not know is the Law, the Force or Forces, which have been used as the instrument of that Purpose. When Man makes a voltaic Battery, he selects materials which have properties and relations with each other previously ascertained— metals worked out of natural ores, acids distilled out of other natural substances; and he puts these together in such fashion as he knows will generate the mysterious Force which he desires to evoke and to employ. But how can such a machine be made out of the tissues of a fish? Well may Mr. Darwin say, "It is impossible to conceive by what steps these wondrous organs have been produced."[1] We see the Purpose—that a special apparatus should be prepared, and we see that it is effected by the production of the machine required; but we have not the remotest notion of the means employed. Yet we can see so much as this, that here again other laws, belonging altogether to another department of Nature—laws of organic growth—are made subservient to a very definite and very peculiar Purpose. The paramount facts

[1] "Origin of Species," p. 192, 1st edition.

disclosed by Science, however, in this case, are these :—first, the adaptation of the animal tissues to form a battery; and, secondly, the Purpose or function of the apparatus, when made, to discharge electric shocks.

There is indeed one objection to this method of conception, which would be a fatal objection if it could be consistently maintained. But all the strength of this objection lies in the obscure terrors which a very long word is sometimes capable of inspiring. This word is "Anthropomorphism." Purpose and Design, it is said, is a human conception. Unquestionably it is, and so is all knowledge in every form. We can never stand outside ourselves. We can never get behind or above our own methods of conception. The human mind can know nothing, and can think of nothing except in terms of its own capacities of thought. But if this be fatal to our knowledge of any of the meanings in creation, it must be equally fatal to our having any knowledge of the very existence of a Creator. Once grant it to be true, "that if we are to apply our human standard to the Creator in one direction, we must apply it in all,"[1]—then it will follow that we cannot conceive any Creator unless it be one as weak, and as corrupt, and as ignorant as ourselves. If this be not bad logic, as on the face of it it clearly is, then it is not "Theology" alone which goes by the board. The purest and

[1] Mr. G. H. Lewes, *Fortnightly Review*, July 1867, p. 109.

most naked Theism is equally destroyed. If it can be said with truth that "the Universal Mind is essentially other than the Human Mind,"[1] so that no recognisable relations can exist between them, then that Universal Mind is to us as if it were not. But those who take objection to Anthropomorphism, are not generally prepared to follow it to this extreme conclusion. Mr. Lewes speaks of the sceptical philosophy he supports as "rejecting Atheism"—of Atheism being "an error which it has not maintained,"—of Atheism being not only rash, but "contradictory."[2] But every conception of a "Mind," even though it be described as "Universal," must be in some degree Anthropomorphic. Our minds can think of another mind only as having some powers and properties which in kind are common with our own. Nor is this objection avoided by any of the other methods of conception which are devised to eliminate from the Order of Nature one of the most patent of its facts. The idea of natural forces working "by themselves" is pre-eminently Anthropomorphic. This is undoubtedly the way in which they seem to us to work when we employ them. The idea of those forces having been so co-ordinated at the first as to produce "necessarily" and "by themselves" all the phenomena of Nature—this is an idea essentially formed on those higher efforts

[1] Mr. G. H. Lewes, *Fortnightly Review*, July 1867, p. 109.

[2] Ibid. p. 107.

of human ingenuity in virtue of which "self-acting" machines are made. It is quite true, no doubt, that this is one aspect in which the adjustments and contrivances in Nature present themselves to us. But it does not render this idea more Anthropomorphic, but rather less when we add to it other conceptions—such as the idea of a Mind which is the source of all power, and a Will which is present in all effects. There may be other difficulties in the way of this conception, but not the difficulty of Anthropomorphism. From neither of these conceptions, however, can we eliminate the idea of Purpose and Design.

It is very difficult to divest ourselves of the notion, that whatever happens by way of natural consequence is thereby removed, at least by one degree, from being the expression of Will and the effect of Purpose. We forget that all our own works, not less than the works of Nature, are works done through the means and instrumentality of Law. All that we can effect is brought about by way of natural consequence. All our machines are simply contrivances for bringing natural Forces into operation; and these machines themselves we are able to construct only out of the materials and by application of the laws of Nature. The Steam-engine works by way of natural consequence; so does Mr. Babbage's Calculating Machine; so does the Electric Telegraph;

so does the Solar System. It is true, indeed, that in all human machinery we know by the evidence of sight the ultimate agency to which the machinery is due, whereas in the machinery of Nature the ultimate agency is concealed from sight. But it is the very business and work of Science to rise from the Visible to the Invisible—from what we observe by Sense to what we know by Reason.

And this brings us to the Fifth meaning in which the word Law is habitually used in Science,—a meaning which is indeed well deserving of attention. In this sense, Law is used to designate, not any observed Order of facts,—not any Force to which such Order may be due,—neither yet any combination of Force adjusted to the discharge of function, but—some purely Abstract Idea, which carries up to a higher point our conception of what the phenomena are and of what they do. There may be no phenomena actually corresponding to such Idea, and yet a clear conception of it may be essential to a right understanding of all the phenomena around us. A good example of Law in this sense is to be found in the law which, in the Science of Mechanics, is called the First Law of Motion. The law is, that all Motion is in itself (that is to say, except as affected by extraneous Forces) uniform in velocity, and rectilinear in direction. Thus according to this law a body moving, and not subject to any extraneous Force, would go on moving for ever at the same rate of velocity, and in an exactly straight line.

Now, there is no such motion as this existing on the earth or in the heavens. It is an Abstract Idea of Motion which no man has ever, or can ever, see exemplified. Yet a clear apprehension of this Abstract Idea was necessary to a right understanding and to the true explanation of all the motions which are actually seen. It was long before this idea was arrived at; and for want of it, the efforts of Science to explain the visible phenomena of Motion were always taking a wrong direction. There was a real difficulty in conceiving it, because not only is there no such motion in Nature, but there is no possibility by artificial means of producing it. It is impossible to release any moving body from the impulses of extraneous Force. The First Law of Motion is therefore a purely Abstract Idea. It represents a Rule which never operates as we conceive it, by itself, but is always complicated with other Rules which produce a corresponding complication in result. Like many other laws of the same class, it was discovered, not by looking outwards, but by looking inwards; not by observing, but by thinking. The human mind, in the exercise of its own faculties and powers, sometimes by careful reasoning, sometimes by the intuitions of genius unconscious of any process, is able, from time to time, to reach now one, now another, of those purely Intellectual Conceptions which are the basis of all that is intelligible to us in the Order of the Material World. We look for an ideal order or simplicity in material Law; and the very possibility of exact Science depends upon the fact that such ideal order does actually prevail, and is related to the abstract conceptions of our own intellectual nature. It is in this way that many of the greatest discoveries of Science have been made. Especially have the great pioneers in new paths of discovery been led to the opening of those paths by that fine sense for abstract truths which is the noblest gift of genius. Copernicus, Kepler, and Galileo were all guided in their profound interpretations of visible phenomena by those intuitions which arise in minds finely organised, brought into close relations with the mind of Nature, and highly trained in the exercise of speculative thought. They guessed the truth before they proved it to be true; and those guesses had their origin in Abstract Ideas of the mind which turned out to be ideas really embodied in the Order of the Universe. So constantly has this recurred in the history of Science, that, as Dr. Whewell says, it is not to be considered as an exception, but as the rule.[1]

1 Whewell's "History of the Inductive Sciences," 2nd edition, vol i. p. 434. Speaking of Copernicus, Dr. Whewell says, in

Here again it is very instructive to observe how "Law" in this last sense is dealt with by the Positive Philosophy. Scientific men are accustomed to reckon such Laws as the First Law of Motion among the surest possessions of pure Intellect, and the faculty by which they are conceived among the noblest proofs of its energy and power. Positivism, on the contrary, regards such laws as mere "artifices" of thought, and the Power by which they are conceived not as a Strength, but as an "Infirmity" of Mind.[1] I do not deny that the process by which these Abstractions are attained is a metaphysical process,—that is to say, they are purely mental conceptions. But the process which denies "reality" to these conceptions is also purely a metaphysical process, with this only difference, that it is bad metaphysics instead of good. The analysis which evolves these abstract Laws out of the phenomena of Nature is an analysis which truly co-ordinates the order of those phenomena with an Order

another place: "It is manifest that in this, as in other cases of discovery, a clear and steady possession of abstract Ideas, and an aptitude in comprehending real Facts under these general conceptions, must have been leading characters in the Discoverer's mind." —Vol. i. p. 389.

1 "Science is distinguished from common knowledge by its conscious employment of artifices which our infirmity renders indispensable." Again, "Abstraction is one of the necessary (from infirmity) artifices of research."—Lewes' "Prologue," p. lxxxix.

of Thought. The counter Analysis which pronounces them to be mere artifices of Thought, and "preliminary falsifications of fact," is an attempt to make Reason disbelieve herself, and immerses us at once in the worst kind of Metaphysics—that which has made the name almost opprobrious—even the old Scholastic subtleties of the Nominalistic and the Realistic controversy.

And now having traced the various senses in which Law is used, we can form some estimate on the value of those conclusions of which some men are so boastful and of which other men are so much afraid. We can see how much and how little is really meant when it is said that Law can be traced in all things, and all things can be traced to Law. It is a great mistake to suppose that, in establishing this conclusion, the progress of modern investigation is in a direction tending to Materialism. This may be and always has been the tendency of individual minds. There are men who would stare into the very Burning Bush without a thought that the ground on which they stand must be Holy Ground. It is not now of wood or stone that men make their Idols, but of their own abstract conceptions. Before these, borrowing for them the attributes of Personality, they bow down and worship. Nothing is more common than to find men who may be trusted thoroughly on the facts of their own Science, who cannot be trusted for a moment on the place which those facts assume in the general system of truth. Philosophy must include Science; but Science does not necessarily include Philosophy. There are, and there always have been, some special misconceptions connected with the prosecution of physical research. It is, however, on the surface of things, rather than below it, that the suggestions of Materialism lie thickest to the eye. They abound among the commonest facts which obtrude themselves on our attention in Nature and in human life. When the bursting of some small duct of blood upon the Brain is seen to destroy in a moment the Mind of Man, and to break down all the powers of his Intellect and his Will, we are in presence of a fact whose significance cannot be increased by a million of other facts analogous in kind.

Yet on every fresh discovery of a few more such facts, there is generally some fresh outbreak of old delusions respecting the forms and the Laws of Matter as the supreme realities of the world. But when the new facts have been looked at a little longer, it is always seen that they take their place with others which have been long familiar, and the eternal problems which lie behind all natural phenomena are seen to be unaffected and unchanged. Like the most distant of the Fixed Stars, they have no parallax. The whole orbit of human knowledge shows in them no apparent change of place. No amount of knowledge of the kind which alone physical Science can impart can do more than widen the foundation of intelligent spiritual beliefs. We think that Astronomy and Geology have given to us in these latter days ideas wholly new in respect to Space and Time. Yet, after all, can we express those ideas, or can we indicate the questions they suggest, in any language which approaches in power to the majestic utterances of David and of Job? We know more than they knew of the magnitude of the Heavenly Bodies; but what more can we say than they said of the wonder of them,—of Orion, of Arcturus, and the Pleiades?[1] We know that the earth moves, which they did not know; and we know that the rapid rotation of a globe on its own axis is a means of maintaining the steadiness of that axis in its course through Space. But what effect, except that of increasing its significance, has this knowledge upon the praise which David ascribes to that ultimate Agency which has made the round world so sure "that it cannot be moved?"[2]

And so of other departments of Science. Even the modern idea of Law, of the constancy and therefore the trustworthiness of Natural Forces, has been known, not indeed scientifically but instinctively, to Man since first

[1] Job ix. 9. [2] Ps. xciii. 1.

he made a Tool, and used it as the instrument of Purpose. What has Science added to this idea, except that the same rule prevails as widely as the Universe, and is made subservient in a like manner to Knowledge and to Will? In the enthusiasm awakened by the discovery of some new facts, or of some new forces, and in the freshness with which they impress the idea of such agencies on our minds, we sometimes very naturally exaggerate the length of way along which they carry us towards the great ultimate objects of intellectual desire. We forget altogether that the knowledge they convey is in quality and in kind identical with knowledge already long in our possession, and places us in no new relation whatever to the vast background of the Eternal and the Unseen. Thus it is that the notions of Materialism are perpetually reviving, and are again being perpetually swept away—swept away partly before the Intuitions of the Mind, partly before the Conclusions of the Reason. For there are two great enemies to Materialism,—one rooted in the Affections, the other in the Intellect. One is the power of THINGS HOPED FOR—a power which never dies: the other is the evidence of THINGS NOT SEEN—and this evidence abounds in all we see. In reinforcing this evidence, and in adding to it, Science is doing boundless work in the present day. It is not the extent of our knowledge, but rather the limits of it, that physical research teaches us to see and feel the most. Of course, in so far as its discoveries are really true, its influence must be for good. To doubt this were to doubt that all truth is true, and that all truth is God's.

There are eddies in every stream—eddies where rubbish will collect, and circle for a time. But the ultimate bearing of scientific truth cannot be mistaken. Nothing is more remarkable in the present state of physical research than what may be called the transcendental character of its results. And what is transcendentalism but the tendency to trace up all things to the relation in which they stand to abstract Ideas? And what is this but to bring all physical phenomena nearer and nearer into relation with the phenomena of Mind? The old speculations of Philosophy which cut the ground from Materialism by showing how little we know of Matter, are now being daily reinforced by the subtle analysis of the Physiologist, the Chemist, and the Electrician. Under that analysis Matter dissolves and disappears, surviving only as the phenomena of Force; which again is seen converging along all its lines to some common centre—"sloping through darkness up to God."[1]

Even the writers who have incurred most reasonable suspicion as to the drift of their teaching, give neverthe-

[1] Tennyson's "In Memoriam."

less constant witness to what may be called the purely mental quality of the ultimate results of physical inquiry. It has been said with perfect truth that "the fundamental ideas of modern Science are as transcendental as any of the axioms in ancient philosophy.[1] We have seen that one of the senses in which Law is habitually used is to designate abstract ideas and doctrines of this kind. So far from these doctrines and ideas having a tendency to Materialism, they serve rather to bring inside the strict domain of Science ideas which in the earlier stages of human knowledge lay wholly within the region of Faith or of Belief. For example, the writer of the Epistle to the Hebrews specially declares that it is by Faith that we understand "that the things which are seen were not made of the things which do appear." Yet this is now one of the most assured doctrines of Science,—that invisible Forces are behind and above all visible phenomena, moulding them in forms of infinite variety, of all which forms the only real knowledge we possess lies in our perception of the Ideas they express—of their beauty, or of their fitness,—in short, of their being all the work of "Toil co-operant to an End."

Every natural Force which we call a law is itself invisible—the idea of it in the mind arising by way of

1 Lewes's "Philosophy of Aristotle," p. 66.

necessary inference out of an observed Order of facts. And very often, if not always, in our conception of these Forces, we are investing them with the attributes of Intelligence and of Will at the very moment, perhaps, when we are stumbling over the difficulty of seeing in them the exponents of a Mind which is intelligent and of a Will which is Supreme. The deeper we go in Science, the more certain it becomes that all the realities of Nature are in the region of the Invisible, so that the saying is literally, and not merely figuratively true, that the things which are seen are temporal, and it is only the things which are not seen that are eternal. For example, we never see the phenomena of Life dissociated from Organisation. Yet the profoundest physiologists have come to the conclusion that Organisation is not the cause of Life, but, on the contrary, that Life is the cause of Organisation,—Life being something—a Force of some kind, by whatever name we may call it—which precedes Organisation, and fashions it, and builds it up. This was the conclusion come to by the great anatomist Hunter, and it is the conclusion endorsed in our own day by such men as Dr. Carpenter and Professor Huxley,—men neither of whom have exhibited in their philosophy any undue bias towards either theological or metaphysical explanations. One illustration referred to by these writers is derived from the shells—the beautiful shells—of the animals called the "Foraminifera."[1] No Forms in Nature are more exquisite. Yet they are the work and the abode of animals which are mere blobs of jelly—without parts, without organs—absolutely without visible structure of any kind. In this jelly, nevertheless, there works a "vital Force" capable of building up an Organism of most complicated and perfect symmetry.

But what is a vital Force? It is something which we cannot see, but of whose existence we are as certain as we are of its visible effects—nay, which our reason tells us precedes and is superior to these. We often speak of Material Forces as if we could identify any kind of Force with Matter. But this is only one of the many ambiguities of language. All that we mean by a Material Force is a force which acts upon Matter, and produces in Matter its own appropriate effects. We must go a step further therefore and ask ourselves, What is Force? What is our conception of it? What idea can we form, for example, of the real nature of that Force, the measure of whose operation has been so exactly ascertained—the Force of Gravitation? It is invisible—imponderable—all our words for it are but circumlocutions to express its phenomena or effects.

[1] "The Elements of Comparative Anatomy," (Huxley,) pp. 10, 11.

There are many kinds of force in Nature—which we distinguish after the same fashion—according to their effects or according to the forms of Matter in which they become cognisable to us. But if we trace all our conceptions on the nature of Force to their fountain-head, we shall find that they are formed on our own consciousness of Living Effort—of that force which has its seat in our own vitality, and especially on that kind of it which can be called forth at the bidding of the Will. In saying this I do not mean to borrow from that false philosophy which pretends by the exercise of reason to get behind all the intuitive convictions on which reason rests. It is in this way that men have come to argue on what they call the "reality of an external world." Even if there were no process of reasoning capable of defending that reality, this would not lend a reasonable character to doubts regarding it. Reason must start from some postulate—some primary truths which cannot be denied. But we need not assume the reality of an external world to be one of these. Yet if it be not a first step, it is a second step hardly distinguishable from the first. Self-existence is of course the truth which may be regarded as the first of all, but in the very idea of Self the existence of that which is Not-Self is necessarily involved. In connecting, however, our conceptions of Force with the consciousness of Living Effort in ourselves, we must guard against mistaking analogy for identity, and against confounding together two items of knowledge which are quite distinct. Correlative with the consciousness of Living Effort in ourselves, and inseparable from it, there is the consciousness of Force acting on us, as well as acting in us. And this argument applies equally whether Self be regarded as a perceiving Mind, or as a physical Organism through which Mind perceives. Thus the knowledge of an external world—that is to say, the knowledge of external Force—stands side by side with the knowledge of Self. Nothing can be known except as distinguished from other things; and all things which are distinguishable from each other, are, in a sense, and in the measure of that distinction, known. And so we know the existence both of internal and of external Force. But if we come to ask ourselves farther questions, as to the nature and seat of Material Force, we can only think of it in the terms of the Vital Force exerted by ourselves. If we can ever know anything of the nature of any Force, it ought to be of this one. And yet the fact is that we know nothing. If, then, we know nothing of that kind of Force which is so near to us, and with which our own Intelligence is in such close alliance, much less can we know the ultimate nature of Force in its other forms.

It is important to dwell on this, because both the aversion with which some men regard the idea of the Reign of Law, and the triumph with which some others hail it, are founded on a notion that, when we have traced any given phenomena to what are called Natural Forces, we have traced them farther than we really have. We know nothing of the ultimate nature, or of the ultimate seat of Force. Science, in the modern doctrine of the Conservation of Energy, and the Convertibility of Forces, is already getting something like a firm hold of the idea that all kinds of Force are but forms or manifestations of some one Central Force issuing from some one Fountain-head of Power. Sir John Herschel has not hesitated to say, that "it is but reasonable to regard the Force of Gravitation as the direct or indirect result of a Consciousness or a Will existing somewhere."[1] And even if we cannot certainly identify Force in all its forms with the direct energies of One Omnipresent and all pervading Will, it is at least in the highest degree unphilosophical to assume the contrary—to speak or to think as if the Forces of Nature were either independent of, or even separate from, the Creator's Power.

It follows, then, from these considerations, that whatever difficulty there may be in conceiving of a Will not exercised by a visible Person, it is a difficulty which

1 "Outlines of Astronomy," 5th edition, p. 291.

cannot be evaded by arresting our conceptions at the point at which they have arrived in forming the idea of Laws or Forces. That idea is itself made up out of elements derived from our own consciousness of Personality. This fact is seen by men who do not see the interpretation of it. They denounce as a superstition the idea of any Personal Will separable from the Forces which work in Nature. They say that this idea is a mere projection of our own Personality into the world beyond—the shadow of our own Form cast upon the ground on which we look. And indeed this, in a sense, is true. It is perfectly true that the Mind does recognise in Nature a reflection of itself. But if this be a deception, it is a deception which is not avoided by transferring the idea of Personality to the abstract Idea of Force, or by investing combinations of Force with the attributes of Mind.

We need not be jealous, then, when new domains are claimed as under the Reign of Law—an agency through which we see working everywhere some Purpose of the Everlasting Will. There are many things in Nature of which we do not see the reason; and many other things of which we cannot find out the cause; but there are none from which we exclude the idea of Purpose by success in discovering the cause. It has been said, with perfect truth, by a living naturalist who is of all others most opposed to what he calls Theological explanations in Science, that we may just as well speak of a watch as the abode of a "watch-force," as speak of the organisation of an animal as the abode of a "vital Force."[1] The analogy is precise and accurate. The Forces by which a watch moves are natural Forces. It is the relation of interdependence in which those Forces are placed to each other, or, in other words, the adjustment of them to a particular Purpose, which constitutes the "watch-force;" and the seat of this Force—which is in fact no one Force, but a combination of many Forces—is in the Intelligence which conceived that combination, and in the Will which gave it effect. The mechanisms devised by Man are in this respect only an image of the more perfect mechanism of Nature, in which the same principle of Adjustment is always the highest result which Science can ascertain or recognise. There is this difference, indeed,—that in regard to our works we see that our knowledge of natural laws is very imperfect, and our control over them is very feeble; whereas in the machinery of Nature there is evidence of complete knowledge and of absolute control. The universal rule is, that everything is brought about by way of Natural

[1] Lewes's "Philosophy of Aristotle," p. 87.

Consequence. But another rule is, that all natural consequences meet and fit into each other in endless circles of Harmony and of Purpose. And this can only be explained by the fact that what we call Natural Consequence is always the conjoint effect of an infinite number of elementary Forces, whose action and re-action are under direction of the Will which we see obeyed, and of the Purposes which we see actually attained.

It is, indeed, the completeness of the analogy between our own works on a small scale, and the works of the Creator on an infinitely large scale, which is the greatest mystery of all. Man is under constraint to adopt the principle of Adjustment, because the Forces of Nature are external to and independent of his Will. They may be managed, but they cannot be disobeyed. It is impossible to suppose that they stand in the same relation to the Will of the Supreme; yet it seems as if He took the same method of dealing with them—never violating them, never breaking them, but always ruling them by that which we call Adjustment or Contrivance. Nothing gives us such an idea of the immutability of Laws as this! nor does anything give us such an idea of their pliability to use. How imperious they are, yet how submissive! How they reign, yet how they serve!

CHAPTER III.

CONTRIVANCE A NECESSITY ARISING OUT OF THE REIGN OF LAW—EXAMPLE IN THE MACHINERY OF FLIGHT.

THE necessity of Contrivance for the accomplishment of Purpose arises out of the immutability of Natural Forces. They must be conformed to, and obeyed. Therefore, where they do not serve our purpose directly, they can only be made to serve it by ingenuity and contrivance. This necessity, then, may be said to be the index and the measure of the power of Law. And so, on the other hand, the certainty with which Purpose can be accomplished by Contrivance, is the index and the measure of mental knowledge and resource. It is by wisdom and knowledge that the Forces of Nature—even those which may seem most adverse—are yoked to service. This idea of the relation in which Law stands to Will, and in which Will stands to Law, is familiar to us in the works of Man : but it is less familiar to us as equally holding good in the works of Nature. We feel, sometimes, as if it were an unworthy notion of the Will which works in Nature, to suppose that it should never act except through the use of means. But our notions of unworthiness are themselves often the unworthiest of all. They must be ruled and disciplined by observation of that which is,—not founded on *à priori* conceptions of what ought to be. Nothing is more certain than that the whole Order of Nature is one vast system of Contrivance. And what is Contrivance but that kind of arrangement by which the unchangeable demands of Law are met and satisfied? It may be that all natural Forces are resolvable into some One Force; and indeed in the modern doctrine of the Correlation of Forces, an idea which is a near approach to this, has already entered the domain of Science. It may also be that this One Force, into which all others return again, is itself but a mode of action of the Divine Will. But we have no instruments whereby to reach this last analysis. Whatever the ultimate relation may be between mental and material Force, we can at least see clearly this,—that in Nature there is the most elaborate machinery to accomplish Purpose through the instrumentality of means. It seems as if all that is done in Nature as well as all that is done in art, were done *by knowing how to do it.* It is curious how the language of the great Seers of the Old Testament corresponds with this idea. They uniformly ascribe all the operations of Nature—the greatest and the smallest—to the working of Divine Power. But they never revolt—as so many do in these weaker days—from the idea of this Power working by wisdom and knowledge in the use of means; nor, in this point of view, do they ever separate between the work of first Creation, and the work which is going on daily in the existing world. Exactly the same language is applied to the rarest exertions of power, and to the gentlest and most constant of all natural operations. Thus the saying that "The Lord by wisdom hath founded the Earth; by understanding hath He established the Heavens,"—is coupled in the same breath with this other saying, "By His knowledge the depths are broken up, and the clouds drop down the dew."[1]

Every instance of Contrivance which we can thoroughly follow and understand, has an intense interest—as casting light upon this method of the Divine government, and upon the analogy between the operations of our own minds and the operations of the Creator. Some instances will strike us more than others—and those will strike us most which stand in some near comparison with our own human efforts of ingenuity and contrivance. There is one such instance which I propose to consider in this chapter—the machinery by which a great pur-

[1] Prov. iii. 19, 20.

pose has been accomplished in Nature—a purpose which Man has never been able to accomplish in art, and that is the Navigation of the Air. No more beautiful example can be found, even in the wide and rich domain of Animal Mechanics—none in which we can trace more clearly, too, the mode and method in which laws the most rigorous and exact are used as the supple instruments of Purpose.

"The way of an Eagle in the air" was one of the things of which Solomon said, that "he knew it not." No wonder that the Wise King reckoned it among the great mysteries of Nature! The Force of Gravitation, though its exact measure was not ascertained till the days of Newton, has been the most familiar of all Forces in all ages of Mankind. How, then, in violation of its known effects, could heavy bodies be supported upon the thin air—and be gifted with the power of sustaining and directing movements more easy, more rapid, and more certain than the movements of other animals upon the firm and solid earth? No animal motion in Nature is so striking or so beautiful as the

> "Scythe-like sweep of wings, that dare
> The headlong plunge through eddying gulfs of air."[1]

Nor will the wonder cease when, so far as the

[1] Longfellow's "Wayside Inn—Ser Federigo."

mechanical problem is concerned, the mystery of flight is solved. If we wish to see how material laws can be bent to purpose, we shall study this problem.

In the first place, it is remarkable that the Force which seems so adverse—the Force of Gravitation drawing down all bodies to the earth—is the very Force which is the principal one concerned in flight, and without which flight would be impossible. It is curious how completely this has been forgotten in almost all human attempts to navigate the air. Birds are not lighter than the air, but immensely heavier. If they were lighter than the air they might float, but they could not fly. This is the difference between a Bird and a Balloon. A Balloon rises because it is lighter than the air, and floats upon it. Consequently, it is incapable of being directed, because it possesses in itself no active Force enabling it to resist the currents of the air in which it is immersed, and because, if it had such a force, it would have no fulcrum, or resisting medium against which to exert it. It becomes, as it were, part of the atmosphere, and must go with it where it goes. No Bird is ever for an instant of time lighter than the air in which it flies; but being, on the contrary, always greatly heavier, it keeps possession of a Force capable of supplying momentum, and therefore capable of overcoming any lesser Force, such as the ordinary resistance of the atmosphere, and even of heavy gales of wind. The Law of Gravitation, therefore, is used in the flight of Birds as one of the most essential of the Forces which are available for the accomplishment of the end in view.

The next law appealed to, and pressed into the service, is again a law which would seem an impediment in the way. This is the resisting force of the atmosphere in opposing any body moving through it. In this force an agent is sought and found for supplying the requisite balance to the Force of Gravity. But in order that the resisting force of air should be effectual for this purpose, it must be used under very peculiar conditions. The resisting force of fluids, and of airs or gases, is a force acting equally in all directions, unless special means are taken to give it predominant action in some special direction. If it is a force strong enough to prevent a body from falling, it is also a force strong enough to prevent it from advancing. In order, therefore, to solve the problem of flight, the resisting power of the air must be called into action as strongly as possible in the direction opposite to the Force of Gravity, and as little as possible in any other. Consequently a body capable of flight must present its maximum of surface to the resistance of the air in the perpendicular direction, and its minimum of surface in the horizontal direction. Now, both these conditions are satisfied (1) by the great breadth or length of surface presented to the air perpendicularly in a Bird's expanded wings, and by (2) the narrow lines presented in its shape horizontally, when in the act of forward motion through the air. But something more yet is required for flight. Great as the resisting force of air is, it is not strong enough to balance the Force of Gravity by its mere pressure on an expanded wing—unless that pressure is increased by an appeal to yet other laws—and other properties of its nature. Every sportsman must have seen cases in which a flying Bird has been so wounded as to produce a rigid expansion of the wings. This does not prevent the Bird from falling, although it breaks the fall, and makes it come more or less gently to the ground.

Yet further, therefore, to accomplish flight, another law must be appealed to, and that is the immense elasticity of the air, and the reacting force it exerts against compression. To enable an animal heavier than the air to support itself against the Force of Gravity, it must be enabled to strike the air downwards with such force as to occasion a rebound upwards of corresponding power. The wing of a flying animal must, therefore, do something more than barely balance Gravity. It must be able to strike the air with such violence as to call forth a reaction equally violent, and in the opposite direction. This is the function assigned to the powerful muscles by which the wings of Birds are flapped with such velocity and strength. We need not follow this part of the problem further, because it does not differ in kind from the muscular action of other animals. The connexion, indeed, between the Wills of animals and the mechanism of their frame, is the last and highest problem of all in the mechanics of Nature; but it is merged and hid for ever in the one great mystery of Life. But so far as this difficulty is concerned, the action of an Eagle's wing is not more mysterious than the action of a Man's arm. There is a greater concentration of muscular power in the organism of Birds than in most other animal frames; because it is an essential part of the problem to be solved in flight, that the engine which works the wings should be very strong, very compact, of a special form, and that, though heavier than the air, it should not have an excessive weight. These conditions are all met in the power, in the outline, and in the bulk of the pectoral muscles which move the wings of Birds. Few persons have any idea of the force expended in the action of ordinary flight. The pulsations of the wing in most Birds are so rapid that they cannot be counted. Even the Heron seldom flaps its wings at a rate of less than from 120 to 150 strokes in a minute. This is counting only the downward strokes, preparatory to each one of which there must be an upward stroke also; so that there are from 240 to 300 separate movements per minute. Yet the Heron is remarkable for its slow and heavy flight, and it is difficult to believe, until one has timed the pulsations with a watch, that they have a rapidity approaching to two in a second. But this difficulty is an index to the enormous comparative rapidity of the faster-flying Birds. Let any one try to count the pulsations of the wing in ordinary flight of a Pigeon, or of a Blackcock, or of a Partridge, or, still more, of any of the diving sea-fowl. He will find that though, in the case of most of these Birds, the quickness of sight enables him to see the strokes separate from each other, it is utterly impossible to count them; whilst in some Birds, especially in the Divers, as well as in the Pheasant and Partridge tribe, the velocity is so great that the eye cannot follow it at all, and the vibration of the wings leaves only a blurred impression on the eye.

Our subject here, however, is not so much the amount of vital force bestowed on Birds, as the mechanical laws which are appealed to in order to make that force effective in the accomplishment of flight. The elasticity of the air is the law which offers itself for the counteraction of gravity. But, in order to make it available for this purpose, there must be some great force of downward blow in order to evoke a corresponding rebound in the opposite, or upward direction. Now, what is the nature of the implement required for striking this downward blow? There are many conditions it must fulfil. First, it must be large enough in area to compress an adequate volume of air; next, it must be light enough in substance not to add an excess of weight to the already heavy body of the Bird; next, it must be strong enough in frame to withstand the pressure which its own action on the air creates. The first of these conditions is met by an exact adjustment of the size or area of the wing to the size and weight of the Bird which it is to lift. The second and the third conditions are both met by the provision of a peculiar substance, feathers, which are very light and very strong; whilst the only heavy parts of the framework, namely, the bones in which the feathers are inserted, are limited to a very small part of the area required.

But there is another difficulty to be overcome—a difficulty opposed by natural laws, and which can only be met by another adjustment, if possible more ingenious and beautiful than the rest. It is obvious that if a Bird is to support itself by the downward blow of its wings upon the air, it must at the end of each downward stroke lift the wing upwards again, so as to be ready for the next. But each upward stroke is in danger of neutralising the effect of the downward stroke. It must be made with equal velocity, and if it required equal force, it must produce equal resistance,—an equal rebound from the elasticity of the air. If this difficulty were not evaded somehow, flight would be impossible. But it is evaded by two mechanical contrivances, which, as it were, triumph over the laws of aërial resistance by conforming to them. One of these contrivances is, that the upper surface of the wing is made convex, whilst the under surface is concave. The enormous difference which this makes in atmospheric resistance is familiarly known to us by the difference between the effect of the wind on an umbrella which is exposed to it on the under or the upper side. The air which is struck by a concave or hollow surface is gathered up, and prevented from escaping; whereas the air struck by a convex or bulging surface escapes readily on all sides, and comparatively little pressure or resistance is produced. And so, from the convexity of the upper surface of a Bird's wing, the upward stroke may be made with comparatively trifling injury to the force gained in the downward blow.

But this is only half of the provision made against a consequence which would be so fatal to the end in view. The other half consists in this—that the feathers of a Bird's wing are made to *underlap each other*, so that in the downward stroke the pressure of the air closes them upwards against each other, and converts the whole series of them into one connected membrane, through which there is no escape; whilst in the upward stroke the same pressure has precisely the reverse effect —it opens the feathers, separates them from each other, and converts each pair of feathers into a self-acting valve, through which the air rushes at every point. Thus the same implement is changed in the fraction of a second from a close and continuous membrane which is impervious to the air, into a series of disconnected joints through which the air passes without the least resistance—the machine being so adjusted that when pressure is required the maximum of pressure is produced, and, when pressure is to be avoided, it is avoided in spite of rapid and violent action.

This, however, exhausts but a small part of the means by which Law is made to do the work of Will in the machinery of flight. It might easily be that violent and rapid blows, struck downwards against the elastic air, might enable animals possessed of such power to lift themselves from the ground and nothing more. There is a common toy which lifts itself in this manner from the force exerted by the air in resisting, and reacting upon little vanes which are set spinning by the hand. But the toy mounts straight up, and is incapable of horizontal motion. So, there are many structures of wing which might enable animals to mount into the air, but which would not enable them to advance or to direct their flight. How, then, is this essential purpose gained? Again we find an appeal made to natural laws, and advantage taken of their certainty and unchangeableness.

The power of forward motion is given to Birds, first by the direction in which the whole wing feathers are set, and next by the structure given to each feather in itself. The wing feathers are all set backwards,—that is, in the direction opposite to that in which the Bird moves; whilst each feather is at the same time so constructed as to be strong and rigid toward its base, and extremely flexible and elastic towards its end. On the other hand, the front of the wing, along the greater part of its length, is a stiff hard edge, wholly unelastic and unyielding to the air. The anterior and posterior webs of each feather are adjusted on the same principle. The consequence of this disposition of the parts as a whole, and of this construction of each of the parts, is, that the air which is struck and compressed in the hollow of the wing, being unable to escape *through* the wing, owing to the closing upwards of the feathers against each other, and being also unable to escape *forwards* owing to the rigidity of the bones and of the quills in that direction, finds its easiest escape *backwards*. In passing backwards it lifts by its force the elastic ends of the feathers;

and thus whilst effecting this escape, in obedience to the law of action and reaction, it communicates, in its passage along the whole line of both wings, a corresponding push forwards to the body of the Bird. By this elaborate mechanical contrivance the same volume of air is made to perform the double duty of yielding pressure enough to sustain the Bird's weight against the Force of Gravity, and also of communicating to it a forward impulse. The Bird, therefore, has nothing to do but to repeat with the requisite velocity and strength its perpendicular blows upon the air, and by virtue of the structure of its wings the same blow both sustains and propels it.[1]

The truth of this explanation of the mechanical theory of flight may be tested in various ways. In the first place it is quite visible to the eye. In many birds flying straight to us, or straight from us, the effect of aërial resistance in bending upwards the ends of the quill feathers is very conspicuous. The flight of the common Rook affords an excellent example—where the Bird is seen foreshortened. In Eagles the same effect is very

[1] The upward stroke has no sustaining power, but has considerable propelling power; because some air, failing to escape between the feathers, must always pass along the convex surface of the wing, and, escaping backwards, must exert upon the ends of the quills a similar reactive force to that which is exerted in the downward stroke.

marked—the wing tips forming a sharp upward curve. I have seen it equally obvious in that splendid Bird the Gannet, or Solan Goose; and when we recollect the great weight which those few quill feathers are thus seen sustaining, we begin to appreciate the degree in which lightness, strength, and imperviousness to the passage of air are combined in this wonderful implement of flight.

But perhaps the simplest test of the action and re-action of the air and the wing feathers in producing forward motion is an actual experiment. If we take in the hand the stretched wing of a Heron, which has been dried in that position, and strike it quickly downwards in the air, we shall find that it is very difficult indeed to maintain the perpendicular direction of the stroke, requiring, in fact, much force to do so; and that if we do not apply this force, the hand is carried irresistibly *forward*, from the impetus in that direction which the air communicates to the wing in its escape backwards from the blow.

Another test is one of reasoning and observation. If the explanation now given be correct, it must follow that since no Bird can flap its wings in any other direction than the vertical—*i.e.* perpendicular to its own axis (which is ordinarily horizontal)—and as this motion has been shown to produce necessarily a forward motion, *no Bird can ever fly backwards.* Accordingly no

Bird ever does so—no man ever saw a Bird, even for an instant, fly tail foremost. A Bird can, of course, allow itself to fall backwards by merely *slowing* the action of its wings so as to allow its weight to overcome their sustaining power; and this motion may sometimes give the appearance of flying backwards,—as when a Swift drops backwards from the eaves of a house, or when a Humming Bird allows itself to drop in like manner from out of the large tubular petals of a flower. But this backward motion is due to the action of gravity, and not to the action of the Bird's wing. In short, it is falling downwards, not flying backwards. Nay, more, if the theory of flight here given be correct, it must equally follow that even standing still, which is the easiest of all things to other animals, must be very difficult, if not altogether impossible, to a Bird when flying. This also is true in fact. To stand still in the air is not indeed impossible to a flying Bird, for reasons to be presently explained, but it is one of the most difficult feats of *wingmanship*,—a feat which many Birds, not otherwise clumsy, can never perform at all, and which is performed only by special exertion, and generally for a very short time, by those Birds whose structure enables them to be adepts in their glorious art.

It cannot be too often repeated—because misconception on this point has been the cardinal error in human attempts to navigate the air—that in all the beautiful evolutions of birds upon the wing, it is weight, and not buoyancy, which makes those evolutions possible. It supplies them, so to speak, with a store of Force which is constant, inexhaustible, inherent in the very substance of themselves, and entirely independent of any muscular exertion. All they have to do is to give direction to that internal Force, by acting on the external Force of aërial currents, through the contraction and expansion of the implements which have been given them for that purpose. Those who have watched the flight of Birds with any care, must have observed that when once they have attained a certain initial velocity and a certain elevation, by rapid and repeated strokes upon the air, they are then able to fly with comparatively little exertion, and very often to pursue their course for long distances without any flapping whatever of the wings. The contrast between the violent efforts required for the first acquisition of the initial velocity, and the perfect ease with which flight is performed after it has been acquired, is a contrast described by Virgil in lines of incomparable beauty:—

> "Qualis speluncâ subito commota columba,
> Cui domus et dulces latebroso in pumice nidi,
> Fertur in arva volans, plausumque exterrita pennis

> Dat tecto ingentem ; mox, aëre lapsa quieto,
> Radit iter liquidum, celeres neque commovet alas."
>
> *Æn.* lib. v. 213–17.

Still more remarkable, as showing the power and the value of weight in flight, is the fact that Birds are able to resume rapid and easy motion not only as the result of a previously-acquired momentum, but after "soaring" in an almost perfectly stationary position. Nothing, for example, is more common than to see Sea Gulls, and some large species of Hawks, "soaring" one moment (that is, all the forces bearing on the Bird brought to an equilibrium, and all motion brought consequently to nearly a perfect standstill), and the next moment sailing onwards in rapid and apparently effortless progression. Now, how is this effect produced? If we only think of it, the question ought rather to be, How is it ever prevented? The soaring is a much more difficult thing to do than the going onwards. It cannot be done at all in a perfectly still atmosphere. It can only be done when there is a breeze of sufficient strength. Gravity is ceaselessly acting on the Bird to pull it downwards: and downwards it must go, unless there is a countervailing Force to keep it up. This force is the force of the breeze striking against the vanes of the wings. But in order to bring these two forces to nearly a perfect balance, and so to "soar," the Bird must expand or contract its wings exactly to the right size, and hold them exactly at the right angle. The slightest alteration in either of these adjustments produces instantly an upsetting of the balance, and of course a resulting motion. The exact direction of that motion will depend on the degree in which the wing is contracted, and the degree in which its angle to the wind is changed. If the wing is very much contracted, and at the same time held off from the wind, that motion will be steeply downwards. Accordingly this is the action of a Hawk when it swoops upon its prey from a great height above it. I have seen a Merlin dash down from a great distance with its wings so closed as to seem almost wholly folded. The Gannet in diving for fish does not close its wings at all, but turning them and the whole axis of its body into the perpendicular, and thus allowing its great weight to act without any counteraction, dashes itself into the sea with foam. But every variety of forward motion is attained by different degrees of contraction and exposure, according to the strength of the breeze with which the Bird has to deal. The limit of its velocity is the limit of its momentum, and the limit of its momentum is the limit of its weight. The lightness of a Bird is therefore a limit to its velocity. The heavier a Bird is, the greater is its possible velocity of flight—because the greater is the store of Force—or to use the language of modern physics, the greater is the quantity of "potential energy" which, with proper implements to act upon aërial resistance, it can always convert into upward, or horizontal, or downward motion, according to its own management and desires.

It will be at once seen from this view of the forces concerned in flight, that the common explanation of Birds being assisted by air-cells for the inhalation and storage of heated air, must not only be erroneous, but founded on wholly false conceptions of the fundamental mechanical principles on which flight depends. If a Bird could inhale enough warm air to make it buoyant, its power of flight would be effectually destroyed. It would become as light as a Balloon, and consequently as helpless. If, on the other hand, it were merely to inflate itself with a small quantity of hot air insufficient to produce buoyancy, but sufficient to increase its bulk, the only effect would be to expose it to increased resistance in cleaving the air. It is true, indeed, that the bones of Birds are made more hollow and lighter than the bones of Mammals, because Birds, though requiring weight, must not have too much of it. It is true, also, that the air must have access to these hollows, else they would be unable to resist atmospheric pressure. But it is no part whatever of the plan or intention of the structure of Birds, or of any part of that structure, to afford balloon-space for heated air with a view to buoyancy.

And here, indeed, we open up a new branch of the same inquiry, showing, in new aspects, how the universality and unchangeableness of all natural laws are essential to the use of them as the instruments of Will; and how by being played off against each other they are made to express every shade of thought, and the nicest change of purpose. The movement of all flying animals in the air is governed and determined by Forces of muscular power, and of aërial resistance and elasticity, being brought to bear upon the Force of Gravity, whereby, according to the universal laws of motion, a direction is given to the animal which is the resultant, or compromise, between all the Forces so employed. Weight, as we have seen, is one of these Forces—absolutely essential to that result, and no flying animal can ever for a moment of time be buoyant, or lighter than the air in which it is designed to move. But it is obvious that, within certain limits, the proportion in which these different Forces are balanced against each other admits of immense variety. The limits of variation can easily be specified. Every flying animal must have muscular power great enough to work its own size of wing: that size of wing must be large enough to act upon a volume of air sufficient to lift the animal's whole weight: lastly, and consequently, the

THE SWIFT.

weight must not be too great, or dispersed over too large a bulk. But within these limits there is room for great varieties of adjustments, having reference to corresponding varieties of purpose. To some Birds the air is almost their perpetual home—the only region in which they find their food—a region which they never leave, whether in storm or sunshine, except during the hours of darkness, and the yearly days which are devoted to their nests. Other Birds are mainly terrestrial, and never betake themselves to flight except to escape an enemy, or to follow the seasons and the sun. Between these extremes there is every possible variety of habit. And all these have corresponding varieties of structure. The Birds which seek their food in the air have long and powerful wings, and so nice an adjustment of their weight to that power and to that length, that the faculty of self-command in them is perfect, and their power of direction so accurate that they can pick up a flying gnat whilst they are passing through the air at the rate of more than a hundred miles an hour. Such especially are the powers of some species of the Swallow tribe, one of which, the common Swift, is a creature whose wonderful and unceasing evolutions seem part of the happiness of summer and of serene and lofty skies.[1]

[1] For the form of the wing in this remarkable bird, see the beautiful drawing here engraved from the pencil of Mr. Wolf.

There are other Birds in which the wing has to be adapted to the double purpose of swimming, or rather of diving, and of flight. In this case, a large area of wing must be dispensed with, because it would be incapable of being worked under water. Consequently in all diving Birds the wings are reduced to the smallest possible size which is consistent with retaining the power of flight at all; and in a few extreme Forms, the power of flight is sacrificed altogether, and the wing is reduced to the size, and adapted to the function, of a powerful fin. This is the condition of the Penguins. But in most genera of swimming Birds, both purposes are combined, and the wing is just so far reduced in size and stiffened in texture as to make it workable as a fin under water, whilst it is still just large enough to sustain the weight of the Bird in flight. And here again we have a wonderful example of the skill with which inexorable mechanical laws are subordinated to special purpose. It is a necessary consequence of the area of the wing being so reduced, in proportion to the size of the Bird, that great muscular power must be used in working it, otherwise the Force of Gravity could not be overcome at all. It is a farther consequence of this proportion of weight to working power, that there must be great momentum and therefore great velocity of flight. Accordingly this is the fact with all the oceanic diving Birds. They have vast distances to go, following shoals of fish, and moving from their summer to their winter haunts. They all fly with immense velocity, and the wing-strokes are extremely rapid. But there is one quality which their flight does not possess—because it is incompatible with their structure, and because it is not required by their habits—they have no facility in evolutions, no delicate power of steering; they cannot stop with ease, nor can they resume their onward motion in a moment. They do not want it: the trackless fields of ocean over which they roam are broad, and there are no obstructions in the way. They fly in straight lines, changing their direction only in long curves, and lighting in the sea almost with a tumble and a splash. Their rising again is a work of great effort, and generally they have to eke out the resisting power of their small wings, not only by the most violent exertion, but by rising against the wind, so as to collect its force as a help and addition to their own.

And now, again, we may see all these conditions changed where there is a change in the purpose to be served. There is another large class of oceanic Birds whose feeding ground is not under water, but on the surface of the sea. In this class all those powers of flight which would be useless to the Divers are absolutely required, and are given in the highest perfection, by the enlistment of the same mechanical laws under different conditions. In the Gulls, the Terns, the Petrels, and in the Fulmars, with the Albatross as their typical Form, the mechanism of flight is carried through an ascending scale, to the highest degrees of power, both as respects endurance and facility of evolution.

The mechanical laws which are appealed to in all these modifications of structure require adjustments of the finest kind; and some of them are so curious and so beautiful that it is well worth following them a little further in detail.

There are two facts observable in all Birds of great and long-sustained powers of flight:—the first is, that they are always provided with wings which are rather long than broad, sometimes extremely narrow in proportion to their length; the second is, that the wings are always sharply pointed at the ends. Let us look at the mechanical laws which absolutely require this structure for the purpose of powerful flight, and to meet which it has accordingly been devised and provided.

One law appealed to in making wings rather long than broad is simply the law of leverage. But this law has to be applied under conditions of difficulty and complexity, which are not apparent at first sight. The body to be lifted is the very body that must exert the lifting power. The Force of Gravity, which has to be resisted, may be said to be sitting side by side, occupying the same particles of matter, with the Vital Force which is to give it battle. Nay, more, the one is connected with the other in some mysterious manner which we cannot trace or understand. A dead Bird weighs as much as a living one. Nothing which our scales can measure is lost when the Vital Force is gone. It is The Great Imponderable. Nevertheless, vital forces of unusual power are always coupled with unusual mass and volume in the matter through which they work. And so it is that a powerful Bird must always also be comparatively a heavy Bird. And then it is to be remembered that the action of gravity is constant and untiring. The Vital Force, on the contrary, however intense it may be, is intermitting and capable of exhaustion. If, then, this Force is to be set against the Force of Gravity, it has much need of some implement through which it may exert itself with mechanical advantage as regards the particular purpose to be attained. Such an implement is the lever—and a long wing is nothing but a long lever. The mechanical principle, or law, as is well known, is this,—that a very small amount of motion, or motion through a very small space, at the short end of a lever, produces a great amount of motion, or motion through a long space, at the opposite or longer end. This action requires indeed a very intense force to be applied at the shorter end, but it applies that force with immense advantage for the purpose in view: because the motion which is transmitted to the end of a long wing is a motion acting at that point through a long space, and is therefore equivalent to a very heavy weight lifted through a short space at the end which is attached to the body of the Bird. Now this is precisely what is required for the purpose of flight. The body of a Bird does not require to be much lifted by each stroke of the wing. It only requires to be sustained; and when more than this is needed—as when a Bird first rises from the ground, or from the sea, or when it ascends rapidly in the air—greatly increased exertion—in many cases, very violent exertion—is required.[1] And then it is to be remembered that long wings economise the vital force in another way. When a strong current of air strikes against the wings of a Bird, the same sustaining effect is produced as when the wing strikes

[1] The Albatross, when rising from tne sea, is described ("Ibis," July 1865) as "stretching out his neck, and with great exertion of his wings, running along the top of the water for seventy or eighty yards, until at last having got sufficient impetus, he tucks up his legs, and is once more fairly launched into the air." The contrast here described between the violent exertion required in first rising, and the perfect ease of flight after this first momentum has been acquired, is a striking illustration of the true mechanical principles of flight.

against the air. Consequently Birds with very long wings have this great advantage, that with pre-acquired momentum, they can often for a long time fly without flapping their wings at all. Under these circumstances, a Bird is sustained very much as a boy's kite is sustained in the air. The string which the boy holds, and by which he pulls the kite downwards with a certain force, performs for the kite the same offices which its own weight and balance and momentum perform for the Bird. The great long-winged oceanic Birds often appear to float rather than to fly. The stronger is the gale, their flight, though less rapid, is all the more easy—so easy indeed as to appear buoyant; because the blasts which strike against their wings are enough to sustain the bird with comparatively little exertion of its own, except that of holding the wing vanes stretched and exposed at proper angles to the wind. And whenever the onward force previously acquired by flapping becomes at length exhausted, and the ceaseless inexorable Force of Gravity is beginning to overcome it, the Bird again rises by a few easy and gentle half-strokes of the wing. Very often the same effect is produced by allowing the Force of Gravity to act, and when the downward momentum has brought the Bird close to the ground or to the sea, that force is again converted into an ascending impetus by a change in the angle at which the wing is exposed to the wind. This is a constant action with all the oceanic Birds. Those who have seen the Albatross have described themselves as never tired of watching its glorious and triumphant motion :—

> "Tranquil its spirit seemed, and floated slow;
> Even in its very motion there was rest."[1]

Rest—where there is nothing else at rest in the tremendous turmoil of its own stormy seas! Sometimes for a whole hour together this splendid Bird will sail or wheel round a ship in every possible variety of direction without requiring to give a single stroke to its pinions. Now, the Albatross has the extreme form of this kind of wing. Its wings are immensely long—about fourteen or fifteen feet from tip to tip—and almost as narrow in proportion as a riband.[2] Our common Gannet is an excel-

[1] Professor Wilson's Sonnet, "A Cloud," &c.

[2] The mechanical principle involved in the sufficiency of very narrow wings has, I believe, been adequately explained in a very ingenious paper read before the *Aëronautical Society*, by Mr. F. H. Wenham, C.E. It is the same mechanical principle which accounts for the narrow blades of a Screw Propeller having a resisting force as great as would be exerted upon the whole area of rotation by a solid Disc. In the case of a flat body, such as the wing of a bird, being propelled edgeways through the air, nearly the whole resisting and sustaining force is exerted upon the first few inches of the advancing surface.

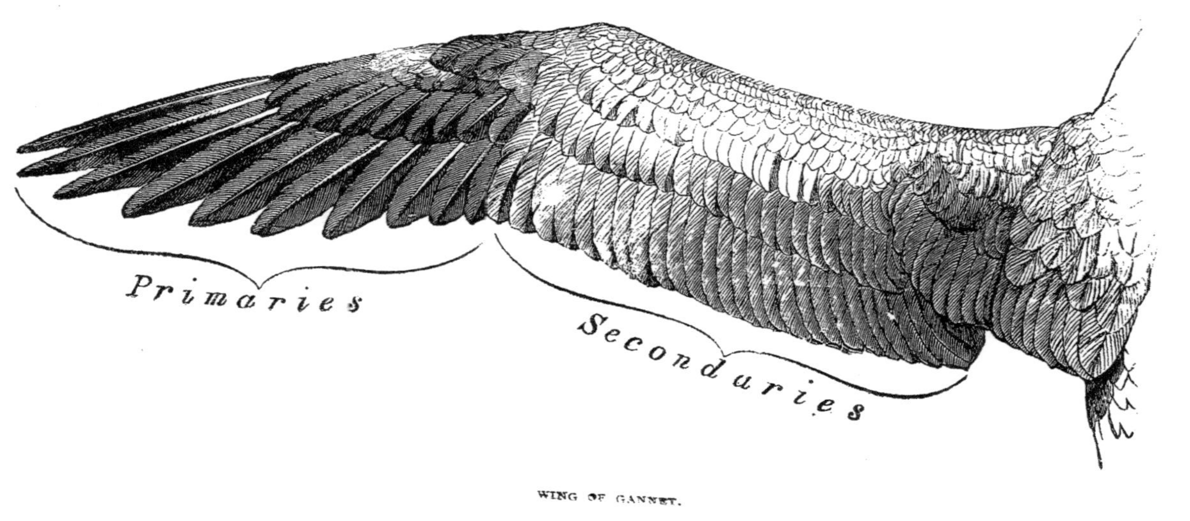

WING OF GANNET.

lent, though a more modified, example of the same kind of structure. On the other hand, Birds of short wings, though their flight is sometimes very fast, are never able to sustain it very long. The muscular exertion they require is greater, because it does not work to the same advantage. Most of the Gallinaceous Birds (such as the common Fowl, Pheasants, Partridges, &c.) have wings of this kind; and some of them never fly except to escape an enemy, or to change their feeding-ground.

The second fact observable in reference to Birds of easy and powerful flight—namely, that their wings are all sharply pointed at the end—will lead us still further into the niceties of adjustment which are so signally displayed in the machinery of flight.

The feathers of a Bird's wing have a natural threefold division, according to the different wing-bones to which they are attached. The quills which form the end of the wing are called the Primaries; those which form the middle of the vane are called the Secondaries; and those which are next the body of the Bird are called the Tertiaries. The motion of a Bird's wing increases from its minimum at the shoulder-joint to its maximum at the tip. The primary quills which form the termination of the wing are those on which the chief burden of flight is cast. Each feather has less and less weight to bear, and less and less force to exert, in proportion as it lies nearer the body of the Bird; and there is nothing more beautiful in the structure of a wing than the perfect gradation in strength and stiffness, as well as in modification of form, which marks the series from the first of the Primary quills to the last and feeblest of the Tertiaries.[1] Now, the sharpness or roundness of a wing at the tip depends on the position which is given to the *longest* Primary quill. If the first, or even the second, primary is the longest, and all that follow are considerably shorter, the wing is necessarily a pointed wing, because the tip of a single quill forms the end; but if the third or fourth Primary quills are the longest, and the next again on both sides are only a little shorter, the wing becomes a round-ended wing. Round-ended wings are also almost always open-ended—that is to say, the tips of the quills do not touch each other, but leave interspaces at the end of the wing, through which, of course, a good deal of air escapes. Since each single quill is formed on the same principle as the whole wing—that is, with the anterior margin stiff and the posterior margin yielding—this escape is not

[1] I owe to the accurate pencil of Mr. J. Wolf the accompanying engraving of the wing of the Golden Plover, a Bird of powerful flight. In this wing the gradation of the feathers is very perfect. It will be observed that the first of the Secondaries, the eleventh feather from the tip of the wing, is marked by a slight variation in the form of the margin.

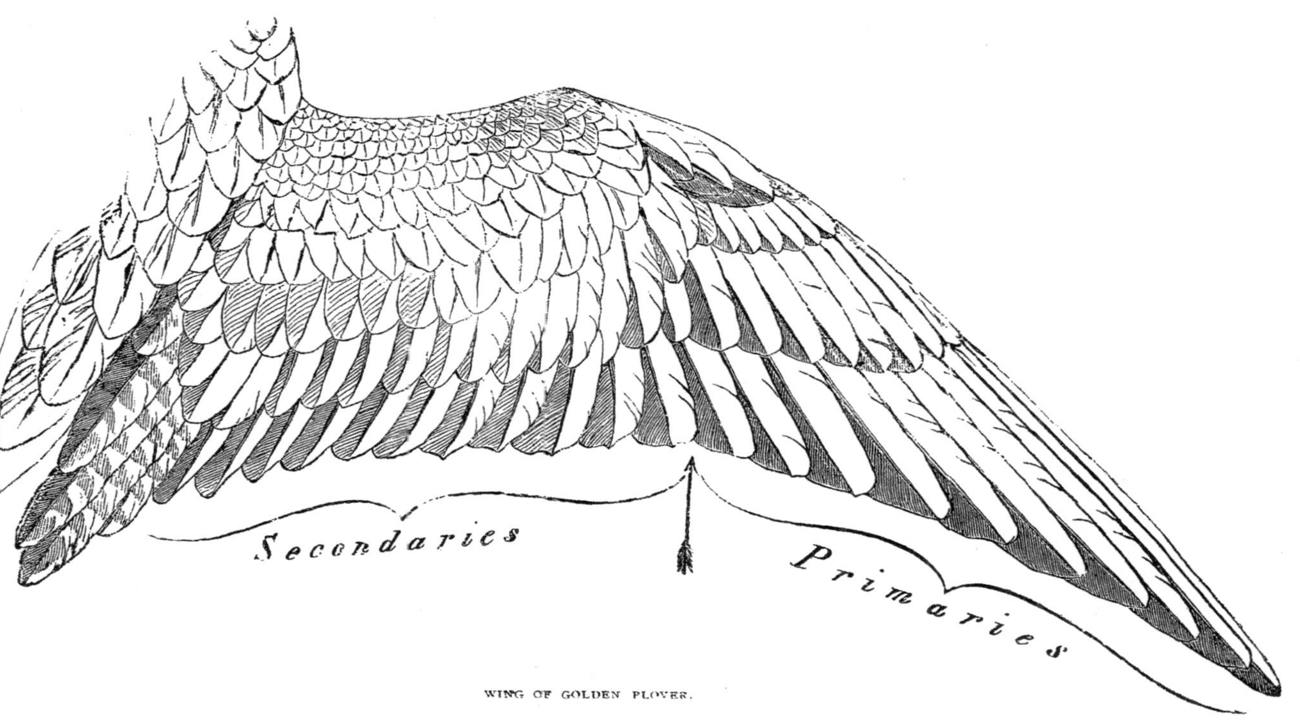

WING OF GOLDEN PLOVER.

useless for progression; but the air acts less favourably for this purpose than when struck by a more compact set of feathers. The common Rook and all the Crows are examples of this. The Peregrine Falcon, the common Swallow, and all Birds of very powerful flight, have been provided with the sharp-pointed structure.[1]

The object of this structure, and the mechanical laws to which it appeals, will be apparent when we recollect what it is on which the propelling power, as distinct from the sustaining power, of a Bird's wing depends. It depends on the reaction of the air escaping *backwards*—that is, in the direction exactly opposite to that of the intended motion of the Bird. Any air which escapes from under the wing, in any other direction, will of course react with less advantage upon that motion. But from under a round wing a good deal of air must necessarily escape *along the rounded end*—that is, in a direction at right angles to the line of intended flight. All the reaction produced by this escape is a reaction which is useless for propulsion. Accordingly, in all Birds to which great velocity of flight is essential, this structure, which is common in other Birds, is carefully avoided.

[1] The illustrations of Mr. Wolf will here again be the best explanation to the reader of the difference between the sharp and the round structure, p. 156.

The Hawks have been classified as "noble" or "ignoble," according to the length and sharpness of their wings: those which catch their prey by velocity of flight having been uniformly provided with the long-pointed structure. The Sparrow-Hawk and the Merlin are excellent examples of the difference. The Sparrow-Hawk, with its comparatively short and blunt wings, steals along the hedgerows and pounces on its prey by surprise; seldom chasing it, except for a short distance, and when the victim is at a disadvantage. And well do the smaller Birds know this habit, and the limit of his powers. Many of them chase and "chaff" the Sparrow-Hawk, when he is seen flying in the open, perfectly aware that he cannot catch them by fast flying. But they never play these tricks with the Merlin. This beautiful little Falcon hunts the open ground, giving fair chase to its quarry by power and speed of flight. The Merlin delights in flying at some of the fastest Birds, such as the Snipe. The longest and most beautiful trial of wingmanship I have ever seen was the chase of a Merlin after a Snipe in one of the Hebrides. It lasted as far as the eye could reach, and seemed to continue far out to sea. In the Merlin, as in all the fastest Falcons, the second quill feather is the longest in the wing; the others rapidly diminish; and the point of the wing looks as sharp as a needle in the air.

There is yet one other power which it is absolutely necessary to some Birds that their wings should enable them to exert: and that is, the power of standing still, or remaining suspended in the air without any forward motion. One familiar example of this is the common Kestrel, which, from the frequent exercise of this power, is called in some counties the "Windhover." The mechanical principles on which the machinery of flight is adapted to this purpose, are very simple. No Bird can exercise this power which is not provided with wings large enough, long enough, and powerful enough to sustain its weight with ease, and without violent exertion. Large wings can always be diminished at the pleasure of the Bird, by being partially folded inwards; and this contraction of the area is constantly resorted to. But a Bird which has wings so small and scanty as to compel it to strike them always at full stretch, and with great velocity in order to fly at all, is incapable of standing still in the air. No man ever saw a Diver or a Duck performing the evolution which the Kestrel may be seen performing every hour over so many English fields. The cause of this is obvious, if we refer to the principles which have already been explained. We have seen that the perpendicular stroke of a Bird's wing has the double effect of both propelling and sustaining. The reaction from such a stroke brings two different forces to bear upon the Bird—one whose direction is upwards, and another whose direction is forwards. How can these two effects be separated from each other? How can the wing be so moved as to keep up just enough of the sustaining force without allowing the propelling force to come into play? The answer to this, although it involves some very complicated laws connected with what mechanicians call the "parallelogram of forces," is practically a simple one. It can only be done by shortening the stroke, and altering the perpendicularity of its direction. Of course, if a Bird, by altering the axis of its own body, can direct its wing-stroke in some degree *forwards*, it will have the effect of stopping instead of promoting progression. But in order to do this, it must have a superabundance of sustaining force, because some of this force is sacrificed when the stroke is off the perpendicular. Hence it follows that Birds so heavy as to require the whole action of their wings to sustain them at all, can never afford this sacrifice of the sustaining force, and, except for the purpose of arresting their flight, can never strike except directly downwards,—that is, directly against the opposing force of gravity. But Birds with superabundant sustaining power, and long sharp wings, have nothing to do but to diminish the length of stroke, and direct it off the perpendicular at such an angle as will bring all the forces bearing upon their body

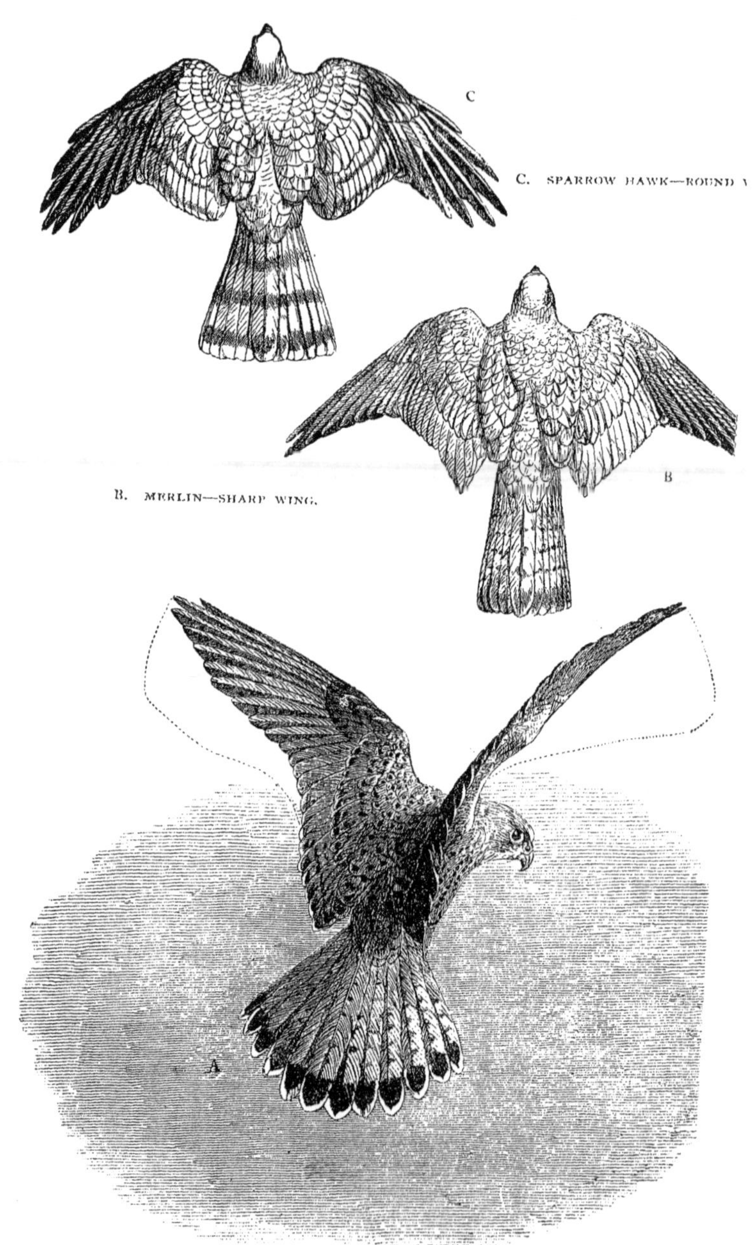

C. SPARROW HAWK—ROUND

B. MERLIN—SHARP WING.

A. KESTREL HOVERING.

to an exact balance, and they will then remain stationary at a fixed point in the air.[1]

They are greatly assisted in this beautiful evolution by an adverse current of air; and it will always be observed that the Kestrel, when hovering, turns *his head to wind*, and hangs his whole body at a greater or less angle to the plane of the horizon. When there is no wind, or very little, the sustaining force is kept up by a short rapid action of the pinions, and the long tail is spread out like a fan to assist in stopping any tendency to onward motion. When there is a strong breeze, no flapping is required at all—the force of the wind supplying the whole force necessary to counteract the force of gravity; and in proportion to the increasing strength of the wind, the amount of vane which must be exposed to it becomes less and less. I have seen a Kestrel stand suspended in a half gale with the wings folded close to the body, and with no visible muscular motion whatever. And so nice is the adjustment of position which is requisite to produce this exact balance of all the forces bearing on the Bird, that the change in that position which again instantly results in a forward motion is very often almost insensible to the eye.

1 Mr. Wolf's illustration of a Kestrel hovering shows accurately the position of the bird when the action is performed in still air.

It is generally a slight expansion of the wings, and a very slight change in the axis of the body.

And here it may be observed that the tails of Birds have not, as is often supposed, any function analogous to the rudder of a ship. Birds which have lost the tail are not thereby rendered incapable of turning. If the steering function had been assigned to Birds' tails, the vane of the tail must have been set, not, as it is, horizontally, but perpendicularly to the line of flight. But a Bird's tail has in flight no lateral motion whatever. It does, indeed, materially assist the Bird in turning, because it serves to *stop the way* of a Bird when it rises or turns in the air to take a new direction. The feathers of the tail are also capable of being depressed unequally,—that is, more at one side than at the other; and when the whole are spread out like the leaves of a fan, this depression at one side is a means whereby the Bird can exert against the air which is passing under it greater muscular pressure upon one side than upon the other, and can thus help the turning action of the wings. With a telescope I have seen this action of the tail very marked in the soaring flight of the Buzzard, when the Bird is wheeling round in spiral circles. The tail contributes also largely to the general balance of the body, which in itself is an important element in the facility of flight. Accordingly, almost all Birds which depend on great ease of evolution in flight—or on the power of stopping suddenly, have largely developed tails. This is the case with all the Birds of prey—with the Kestrel in a conspicuous degree. But there are some exceptions which show that great powers of flight are not always dependent on the possession of a large tail—as, for example, the Swift.

Another explanation has been given of the means by which Birds are able to turn in flight, which is a curious example how preconceived theories founded on false analogies will vitiate our observation of the commonest facts in nature. I do not know of any modern work that gives any account of the theory of flight, which is even tolerably correct. But in most points an admirable account is to be found in the celebrated work of Borelli, "De Motu Animalium." On the question, however, of steerage in flight, he gives a solution which the most ordinary observation is sufficient to contradict. Borelli is quite aware that the tail in Birds has no such function as that which is usually assigned to it, and he points out the true theoretical objection to the possibility of its having any guiding power—viz., its horizontal position, and its immobility in the lateral direction. But the theory which he himself propounds is equally erroneous. It is this,—that Birds deflect their course to the right or to the left, as rowers turn a rowboat—by striking more quickly and more strongly with one wing than with the other.[1] To this theory there are two objections—first, that as matter of fact Birds can turn, and do turn, even to the extent of describing complete circles in the air, without any flapping either of one wing or the other: and secondly, that when Birds do flap and turn at the same time, not the slightest difference in time between the two wing-strokes can ever be detected. The beats of a Bird's two wings are always exactly synchronous. But the first of these two objections is of itself quite sufficient to disprove the theory. No man can have watched even for a moment the flight of the common Swallow, and especially the flight of the Swift, without seeing it perform complete gyrations in the air without any strokes of either wing. The only change which can ever be detected by the eye is a slight elevation on one side

1 Referring to a boat, he says:—"Si remi dexteri lateris celerius quam sinistri aquam retrorsum impellant—semper prora revolvetur versus sinistrum latus; ergo eodem modo dum avis in medio fluido aeris innatat, volando æquilibrata in centro gravitatis ejus, si sola dextra ala deorsum sed oblique flectatur, aerem subjectum impellando versus caudam necessario ad instar navis mox memoratæ, permovetur latus ejus dextrum, quiescente aut tardius moto sinistro latera. Ex quo fit, ut avis pars anterior circa centrum gravitatis ejus revoluta, flectatur versum sinistrum latus." —Borellus, "De Motu Animalium," Pars Prima. Propositio cxcix.

of the whole body, and a slight depression of the other. The depression is always on that side towards which the bird is turning. On the opposite side, that from which the Bird is turning, there is of course a corresponding elevation. Sometimes this is very obvious; but in general it is so slight as to require close observation to detect it. In the Albatross, when sweeping round, the wings are often pointed in a direction nearly perpendicular to the sea.[1] The effect of this, of course, is to expose the two vanes at different angles to the aërial currents—and it must be remembered that in flight the balance of all the forces employed is so extremely fine that the most minute alteration in the degree in which they bear upon each other will produce an immense change in the result. It is not surprising, therefore, that the muscular movements which serve to turn the axis of a flying Bird from one direction to another, are very

1 See a very interesting account of the flight of the Albatross by Captain T. W. Hutton, in the "Ibis" for July 1864. Captain Hutton says: "If he wishes to turn *to the right*, he bends his head and tail slightly upwards, at the same time *raising his left side and lowering the right*, in proportion to the sharpness of the curve he wishes to make, the wings being kept rigid the whole time." This is the only paper I have seen on the flight of birds in which observation of the facts is not vitiated by some false preconceived theory on their cause. Captain Hutton has thoroughly seized the true mechanical principles of flight.

often so extremely minute as generally altogether to elude the sight. But in general terms, it may be said that a Bird turns in flying essentially on the same principle as that on which a Man turns in walking. It is done in both cases by change in the direction of muscular pressure upon a resisting medium. By an exquisite combination of different laws, and by mechanical contrivance in the adjustment of them, it has been given to a Bird to find in the thin and yielding air a medium of resistance against which its own muscular force may act, as firm and as effective as that which Man finds in the solid earth.

The Humming Birds are perhaps the most remarkable examples in the world of the machinery of flight. The power of poising themselves in the air,—remaining absolutely stationary whilst they search the blossoms for insects,—is a power essential to their life. It is a power accordingly which is enjoyed by them in the highest perfection. When they intend progressive flight, it is effected with such velocity as to elude the eye. The action of the wing in all these cases is far too rapid to enable the observer to detect the exact difference between that kind of motion which keeps the Bird at absolute rest in the air, and that which carries it along with such immense velocity. But there can be no doubt that the change is one from a short quick stroke delivered obliquely forward, to a full stroke, more slow, but delivered perpendicularly. This corresponds with the account given by that most accurate ornithological observer, Mr. Gould. He says: "When poised before any object, this action of the wing is so rapidly performed that it is impossible for the eye to follow each stroke, and a hazy semicircle of indistinctness on each side of the Bird is all that is perceptible." There is another fact mentioned by those who have watched their movements most closely which corresponds with the explanation already given—viz., the fact that the axis of the Humming Bird's body when hovering is always *highly inclined*, so much so as to appear almost perpendicular in the air. In other words the wing-stroke, instead of being delivered perpendicularly downwards, which would infallibly carry the body onwards, is delivered at such an angle forwards as to bring to an exact balance the upward, the downward, and the forward forces which bear upon the body of the Bird. Mr. Darwin says, "When hovering by a flower, the tail is constantly shut and expanded like a fan, *the body being kept in a nearly vertical position.*" Mr. Wallace, another accurate observer, describes the Humming Birds as "balancing themselves vertically in the air."

These are a few, and a few only, of the adjustments

required in order to the giving of the power of flight;—adjustments of organic growth to intensity of vital force—of external structure to external work—of shape in each separate feather to definite shape in the series as a whole—of material to resistance—of mass and form to required velocities; adjustments, in short, of law to law, of force to force, and of all to Purpose. So many are these contrivances, so various, so fine, so intricate, that a volume might be written without exhausting the beauty of the method in which this one mechanical problem has been solved. It is by knowledge of unchanging laws that these victories over them seem to be achieved: yet not by knowledge only, except as the guide of Power. For here as everywhere else in Nature, we see the same mysterious need of conforming to imperative conditions, side by side with absolute control over the forces through which this conformity is secured. When any given purpose cannot be attained without the violation of some law, unless by some new power, and some new machinery—the requisite power and mechanism are evolved generally out of old materials, and by modifications of pre-existing forms. There can be no better example of this than a wing-feather. It is a production wholly unlike any other animal growth—an implement specially formed to combine strength with lightness, elasticity, and imperviousness to air. Again, the bones of a Bird's wing are the bones of the Mammalian arm and hand, specially modified to support the feathers. The same purpose is effected by other means in connexion with precisely the same bones in the flying Mammalia—the Bats. In these animals the finger-bones instead of being compressed or soldered together to support feathers, are separated, attenuated, and greatly lengthened to afford attachment to a web or flying membrane which is stretched between them. In other ages of the world there were also flying Lizards. But in all these cases the mechanical principle is the same, and there has been the same ingenious adaptation of material and of force to the universal laws of motion.

On the earth and on the sea Man has attained to powers of locomotion with which, in strength, endurance, and in velocity, no animal movement can compare. But the air is an element on which he cannot travel—an ocean which he cannot navigate. The Birds of heaven are still his envy, and on the paths they tread he cannot follow. As yet! for it is not certain that this exclusion is to be perpetual. His failure has resulted quite as much from his ignorance of natural laws, as from his inability to meet the conditions which they demand. All attempts to guide bodies buoyant in the air must be fruitless. Balloons are mere toys. No flying animal has ever been formed on the principle of buoyancy. Birds and Bats, and Dragons, have been all immensely heavier than the air, and their weight is one of the forces most essential to their flight. Yet there is a real impediment in the way of Man navigating the air—and that is the excessive weight of the only great mechanical moving powers hitherto placed at his disposal. When Science shall have discovered some moving power greatly lighter than any we yet know, in all probability the problem will be solved.[1] But of one thing we may be sure—that if Man is ever destined to navigate the air, it will be in machines formed in strict obedience to the mechanical laws which have been employed by the Creator for the same purpose in flying animals.[2]

[1] The men of Science in France are ahead of the men of Science in England upon this subject. There is a society established in Paris which announces in its very title the true fundamental principle of flight—"Société d'Encouragement pour la Locomotion aérienne au moyen d'Appareils PLUS LOURDS que l'Air." The false principle of Buoyancy is thus eliminated and banished from the question.

[2] I owe to my father (John, 7th Duke of Argyll) my knowledge of the Theory of Flight which is expounded in this chapter. The retired life he led, and the dislike he had of the work of literary composition, confined the knowledge of his views within a comparatively narrow circle. But his love of mechanical science, and his study of the problem during many years of investigation and experiment, made him thoroughly master of the subject. In his devices for testing and illustrating the truth of his Theory, he was chiefly assisted by two very ingenious men, the late Mr. John Hart,

of Glasgow, and the late Mr. Robert Bryson, of Edinburgh. The result of his investigations led him to the opinion that until a lighter moving power than steam is discovered, it will be impossible to construct successfully machines for the navigation of the air. I shall only add, that having made ornithology a favourite pursuit, I have been led during many years to test this theory by close observation of the flight of Birds ; and that from the manner in which it fits into, and explains all the facts, I have been always more and more satisfied of its truth.

CHAPTER IV.

APPARENT EXCEPTIONS TO THE SUPREMACY OF PURPOSE.

YET, as we look at Nature, the fact will force itself upon us that there are structures in which we cannot recognise any use; that there are contrivances which often fail of their effect; and that there are others which appear to be separated from the conditions they were intended to meet, and under which alone their usefulness could arise. Such instances occur in many branches of inquiry; and although in the great mass of natural phenomena the supremacy of Purpose is evident enough, such cases do frequently come across our path as cases of exception—cases in which Law does not seem to be subservient to Will, but to be asserting a power and an endurance of its own.

The degree of importance which may be attached to such cases as a source of real difficulty, will vary with the character of the individual mind, and its capacity of holding by the great lines of evidence which run through the whole Order of Nature. It is with these cases as with the local currents which sometimes obscure the rising and falling of the tides. When watched from hour to hour, the greater law is clearly discernible by well-marked effects: but when watched from minute to minute, that law is not distinct, and there are waves which seem like a rebellion of the sea against the force which is dragging it from the land. The Order of Nature is very complicated, and very partially understood. It is to be expected therefore that there should be a vast variety of subordinate facts, whose relation to each other and to the whole must be a matter of perplexity to us. It is so with the relation in which different known laws of Nature stand to each other; much more must it be so with the far deeper subject of the relation which these laws bear to the Will and the intentions of the Supreme. But as cases of intention frustrated, of structure without apparent purpose, of organs dissociated from function and from the opportunities of use, are sometimes sources of difficulty, it may be well to consider this subject a little nearer. Let us look at it both in the light of abstract reasoning, and also in the light of particular illustration.

In the first place, then, we must remember that results which may appear as exceptions to the attainment of one Purpose may be nothing more than fulfilments of another. This follows from the truth which has been dealt with in a former page,[1] that we are "greatly ignorant," as Bishop Butler says, how far anything in Nature is to be regarded as a means or as an end, and that ultimate or final purposes we can never see. The difficulty hence arising has often been represented as a fundamental objection to the whole doctrine of Intention. But this view is founded on a very great, although a very natural confusion of thought. The perception of Purpose and Intention is inseparable from the perception of Adjustment and Function as these are exhibited in Nature. As such it belongs to Knowledge. It is the perception of a relation between those phenomena and certain well known phenomena of Mind. But to perceive a relation is not necessarily to perceive all that this relation involves. To perceive intention is a very different thing from perceiving all that is intended. Our own human experience should make this distinction familiar to us. Many things we do and many things we contrive are done and contrived with more than one intention. In the light of this experience it is altogether irrational to regard as an exception to the attainment of Purpose in Nature the fact, for example, "that of fifty seeds she often brings but one to bear." It throws no doubt or difficulty in the way of our conviction, for example, that one purpose of seed-

[1] *Ante*, p. 80.

bearing in Plants is the reproduction of their kind, because it appears that another purpose to which that seed-bearing is applied is the support of animal life. The intention with which a grain of wheat is so constituted as to be capable of producing another wheat plant, is not the less in the nature of Purpose because it co-exists with another intention, that the same grain should be capable of sustaining the powers and the enjoyments of Life in the Body and in the Mind of Man. On the contrary, the power possessed by most plants, and by this plant especially, of producing seed in a ratio far beyond that which would be required for one purpose, is the sure indication and the proof that another purpose larger and wider was in view. Yet the seeds of corn which, as seeds, are destroyed when they are converted into bread, may in that aspect be represented and regarded as "failures." In reference to this kind of failure, it has been actually argued that in Nature "the prodigality of waste is far more conspicuous than the wise economy of which so much is said."[1] When applied to the case of the wheat plant the fallacy is apparent, and would probably be admitted. But this is only one example of a class to which an infinite number of other examples in Nature may be

[1] Mr. G. H. Lewes, *Fortnightly Review*, July 1867, p. 100.

referred. There may be, indeed, and there are, innumerable examples where the meaning of like "failures" is not equally evident to us—some which may be involved in utter and hopeless darkness—some which may run up into the great master difficulty—that which we are accustomed to call the "Origin of Evil." But the same argument applies to all. It is not that Purpose and Intention solve all difficulties. But it is that no difficulty in perceiving what may be the purpose and intention of a particular fact can affect the reality and truth of that perception in other cases where no such difficulty exists. Let us now look at the same subject in the light of particular examples.

There is one explanation which, it cannot be doubted, applies to many cases; and this is, the simple explanation that we often mistake the purpose of particular structures in Nature, and connect them with intentions which are not, and never were, the intentions really in view. The best naturalists are liable to such mistakes. A very curious illustration is afforded by an observation of Mr. Darwin, in his "Origin of Species." He says that "if green Woodpeckers alone had existed, and we did not know that there were any black and pied kinds, I daresay we should have thought that the green colour was a beautiful adaptation to hide this tree-frequenting bird from its enemies." Now, this introduces us to a very curious subject, and one as well adapted as any other to illustrate the relation in which Law stands to Purpose in the economy of Nature.

There can be no doubt that the principle of adapted colouring with the effect and for the purpose of concealment, prevails extensively in various branches of the Animal Kingdom. It arises, probably, like all other phenomena, by way of Natural Consequence, out of some combination of forces which are the instruments employed. We have no knowledge what these forces are; but we can imagine them to be worked into a law of assimilation, founded on some such principle as that which photography has revealed. It is true that Man has not yet discovered any process by which the tints of Nature can be transferred, as the most delicate shades of light can be transferred, to surfaces artificially prepared to receive them. Such a process is, however, very probably within the reach even of human chemistry, and it is one which is certainly known in the laboratory of Nature. The Chameleon is the extreme case in which the effect of such a process is proverbially known. Many Fish exhibit it in a remarkable degree, changing colour rapidly in harmony with the colour of the water in which they swim, or of the bottom on which they lie. The law on which such changes depend is very obscure: but it appears to be a natural process, as constant as all other laws are—that is, constant whenever given conditions are brought together. It is possible that the effect may be due to a cause which is well known to be capable of producing somewhat analogous results. Even before the days of Jacob and of Laban, it seems to have been known that through the eyes of the female parent colour can be determined in her young; and although this is certainly not the law which commonly determines colour—operating as it does, so far as we know, seldom, and only in a small degree—it is quite conceivable that, under special conditions, it is capable of being worked as a great power in Nature. But, then, these conditions are not brought together except with a view to purpose. For now let us see how this law, whatever it may be, is regulated and applied.

One thing is certain: assimilated colouring is not applied universally; on the contrary, it is applied very partially. Is it, therefore, applied arbitrarily—at haphazard, or without reference to conditions in which we can trace a reason and a rule? Far from it. The rule appears to be this:—adaptive colouring, as a means of concealment, is never applied (1) to any animal whose habits do not expose it to special danger, or (2) to any animal which is sufficiently endowed with other more effective means of escape.

This is the higher Law of Purpose which governs the

lesser law, whatever it may be, by which assimilative colouring is produced. Now, no man who had observed this higher law could ever fall into the error of supposing that the colour of the Green Woodpecker was given to it as a means of concealment. Few Birds are so invisible as Woodpeckers, because their structure and habits give them other methods of escaping observation, which are most curious and effective. They have few natural enemies but Man; and when in danger of being seen by him, they slip and glide round the bole of a tree or bough on which they may be climbing, with a swift, silent, and cunning motion, and from behind that shelter, with nothing visible but their head, they keep a close watch upon the movements of the enemy. With such sleight of feet, there is no need of lazier methods of concealment.

Accordingly, in this family of Birds, the law of assimilation is withheld from application, and the most violent and strongly contrasted colouring prevails. Jet black, side by side with pure white, and the most brilliant crimsons, are common in the plumage of the Woodpeckers. No birds are more conspicuous in colouring, yet none are more seldom seen. The Green Woodpecker itself, with its yellow tints and crimson hood, contrasts strongly with the bark on which it climbs. The purpose of concealment being effected by other means, gives way to the purpose of beauty or of adornment in the disposition of colours. And in general the same rule applies to all Birds whose life is led among woods and forests. Comparatively inaccessible to Birds of prey, they exhibit every variety of tint, and the principle of invisibility from assimilated colouring is almost unknown.

It must always be remembered, that animals of prey are as much intended to capture their food, as their victims are intended to have some chances and facilities of escape. The purpose here is a double purpose—a purpose not in all cases to preserve life, but to maintain its balance and due proportion. In order to effect this purpose, the means of aggression, and of defence, or of escape, must bear a definite relation to each other both in kind and in degree. When arboreal Birds leave their sheltering trees, they are exposed to the attacks of Hawks, but they have fair opportunities of retreating to their coverts again; and the upward spring of the disappointed Falcon in the air, when his quarry reaches the shelter of trees, tells how effective such a retreat is, and how completely it ends the chase. On the other hand, there is a great variety of Birds whose habitat is the open plain—the desert—the unprotected shore—the treeless moor—the stony mountain-top. These are the favourite hunting-grounds of the Eagles, and the Falcons, and the Hawks. There they have free scope for their great powers of wing, and uninterrupted range for their piercing powers of sight. And it must be remembered, that even the slowest of the Hawks can on such ground capture with ease Birds which, when once on the wing, could distance their pursuer by superior speed, because the Hawk, sweeping over the ground, takes the prey at a disadvantage, pouncing on it before it can get fairly into the air. Birds whose habitat is thus exposed could not maintain their existence at all without special means of concealment or escape. Accordingly it is among such Birds almost exclusively that the law of assimilative colouring prevails. And among them it is carried to a perfection which is wonderful indeed. Every ornithologist will recognise the truth of the observation, that this law prevails chiefly among the Grouse, the Partridges, the Plovers, the Snipes, Woodcocks, Sandpipers, and other kindred families, all of which inhabit open ground. There can be no better examples than the Grouse and the Ptarmigan of our Scottish mountains. The close imitation in the plumage of these Birds of the general tinting and mottling of the ground on which they lie and feed is apparent at a glance, and is best known to those who have tried to see Grouse or Ptarmigan when sitting, and when their position is indicated within a few feet or a few inches by the trembling nostrils and dilated eyeballs of a steady

Pointer-Dog. In the case of the common Grouse, as the ground is nearly uniform in colour throughout the year, the colouring of the Bird is constant also. But in the case of the Ptarmigan, it changes with the changing seasons. The pearly grays which in summer match so exactly with the lichens of the mountain peaks, give place in winter to the pure white which matches not less perfectly with the wreaths of snow.

This is indeed a change which requires for its production the agency of other laws than those merely of reflected light, because the substitution of one entire set of feathers for another of a different colour, twice in every year, implies arrangements which lie deep in the organic chemistry of the Bird. The various genera of Sand-Grouse and Sand-Partridges which frequent the deserts and naked plains of the Asiatic continent, are coloured in exquisite harmony with the ground. Our common Woodcock is another excellent example, and is all the more remarkable as there is one very peculiar colour introduced into the plumage of this Bird which exactly corresponds with a particular stage in the decay of fallen leaves—I mean that in which the browns and yellows of the Autumn rot away into the pale ashy skeletons which lie in thousands under every wood in winter. This colour is exactly reproduced in the feathers of the Woodcock, and so mingled with the dark browns and warm yellows of fresher leaves, that the general imitation of effect is perfect. And so curiously is the purpose of concealment worked out in the plumage of the Woodcock, that one conspicuous ornament of the bird is covered by a special provision from the too curious gaze of those for whose admiration it was not intended. The tail-feathers of the Woodcock can be erected and spread out at pleasure like a fan, and, being tipped on their under surface with white of a brilliant and silvery lustre, set off by contrast with an adjacent patch of velvety black, they then produce a most conspicuous effect. But the same web which on its under surface bears this beautiful but dangerous ornament, is on its upper surface dulled down to a sombre ashy-gray, and becomes as invisible as the rest of the plumage. These are all provisions of Nature, which stand in clear and intelligible relation to the habits of the Bird. It rests all day upon the ground, under trees; and were it not for its ingeniously adapted colouring, it would be peculiarly exposed to destruction. Man is an enemy whose cunning inventions overcome all such methods of protection, and the Woodcock, when in his most rapid flight, is now an easier prey than in older times when sitting on the ground. But before fire-arms had reached the perfection which has enabled us to shoot flying Birds, the colouring of the Woodcock served it in good stead, even against the Lords of the Creation. In old times it required special skill and practice to see Woodcocks on the ground, and the large lustrous black eye which is adapted for night-vision was the one spot of colour which enabled the fowler of a century and a half years ago to detect the bird. Thus Hudibras has it:—

"For fools are known by looking wise,
As men find woodcocks by their eyes."[1]

In Snipes, again, there is a remarkable series of straw-coloured feathers introduced along the back and shoulders, which perfectly imitate the general effect of the bleached vegetable stalks common on the ground which the Bird frequents.

There are other animals in which the principle of imitation with a view to concealment is carried very much farther than the mere imitation of colour, and extends also to form and structure. There are some examples of this in the Class of Insects, so remarkable that it is impossible to look at them without ever fresh astonishment. I refer to some families of the Orthopterous order, and especially to some genera of the *Mantidæ* and *Phasmidæ*. Many species of the genus *Mantis* are wholly modelled in the form of vegetable

[1] "Hudibras to Sidrophel," 79, 80.

growths. The legs are made to imitate leaf-stalks, the body is elongated and notched so as to simulate a twig; the segment of the shoulders is spread out and flattened in the likeness of a seed-vessel; and the large wings are exact imitations of a full-blown leaf, with all its veins and skeleton complete, and all its colour and apparent texture. There is something startling and almost horrible in the completeness of the deception—very horrible it must be to its hapless victims. For in this case the purpose of the imitation is a purpose of destruction, the *Mantis* being a predacious insect, armed with the most terrible weapons, hid under the peaceful forms of the vegetable world. It is the habit of these creatures to sit upon the leaves which they so closely resemble, apparently motionless, but really advancing on their prey with a slow and insensible approach. Their structure disarms suspicion. Wonderful as this structure is, it would be none the less, but all the more wonderful, if it should arise by way of Natural Consequence from some law of development or of growth. It must be a law of which at present we have no knowledge, and can hardly form any conception. But certain it is that here, as in all other cases, the purpose which is actually attained, is attained by a special adaptation of ordinary structure to a special and extraordinary purpose. No new members are given to the *Mantis;* there is no departure from the plan on which all other Insects of the same Order are designed. The body has the same number of segments, the legs are the same in number, and are composed of the same joints; every part of this strange creature which seems like a bit of foliage animated with insect life, can be referred to its corresponding part in the ordinary anatomy of its Class. The whole effect is produced by a little elongation here, a little swelling there, a little dwarfing of one part, a little development of another. The most striking part of the whole imitation—that of the "nervation" of the leaf—is produced by a modification, not very violent, of a structure which belongs to all flying Insects. Their wings are constructed of a thin filmy material stretched upon a framework of stronger substance, as the sails of a windmill are stretched upon a trellis-work of spars. This framework is designed in a great variety of patterns—more elaborate and more beautiful than the tracery of Gothic windows. In the *Mantis* this tracery, instead of being drawn in a mere pattern, is drawn in imitation of the nervature of a leaf. And imitative colouring is added to imitative structure—so that nothing should be wanting to its completeness and success.

It must always be remembered, however, that Contrivance in Nature can never be reduced to a single purpose, and to that alone. Almost every example of it is connected with a number of effects which fit into each other in endless ramifications of adjustment. For example, this imitative structure of the *Mantidæ* serves as well for their own protection from insectivorous birds as for the procuring of their food in the capture of other Insects. And this, which is perhaps the subordinate purpose in the case of the *Mantidæ*, emerges as the main purpose in another family of imitative Insects, the *Phasmidæ*. These last are vegetable feeders, and their imitative structure is, if possible, even more wonderful, as it certainly is more beautiful. In some species the wings are not only made like leaves in form, in structure, and in general colour, but they are tinted at different seasons of the year with the varying colours of spring, of summer, or of autumn. The fundamental green is shaded off into browns, and reds, and yellows, with a few of those crimson touches which are so common in the "Pageant of the year." There is one specimen in the British Museum where the imitative effect is pursued, as it were, into a region of still more minute and curious observation. The general aspect of summer vegetation is much affected by the ravages of insect life. Minute larvæ eat into the cuticle of leaves, and mark them with various spots of bleached or faded colour. Now the specimen of *Phasma* I refer to has its wing covered with spots which exactly imitate this appearance of a larva-eaten leaf. Can it be that this effect is itself produced by a really similar cause—the eating of some larval parasite into the tissue of the wing? If so, the combination of means to the production of so wonderful an effect becomes only the more bewildering in the endless vistas of adjustment which are opened out. And there is another fact connected with these Insects which is as astonishing as any other. It is this—that the idea and purpose of imitation is carried into effect consistently and perseveringly through all the stages of the creature's metamorphoses. The eggs are as perfect imitations of vegetable seeds as the adult insect is of the expanded leaf. In the larval form they are like bits of stalk, or chips or cuttings of leaves.

But although the laws which determine both form and colouring are here seen to be subservient to use, we shall never understand the phenomena of Nature unless we admit that mere ornament or beauty is in itself a purpose, an object, and an end. Mr. Darwin denies this; but he denies it under the strange impression, that to admit it would be absolutely fatal to his own theory on the Origin of Species. So much the worse for his theory, if this incompatibility be true. There is indeed a difference, at least in words, between the doctrine now asserted and the doctrine which Mr.

Darwin denies. What he denies as a purpose in nature is beauty "in the eyes of Man." But this evades the real point at issue. The relation in which natural beauty stands to Man's appreciation of it, is quite a separate question. It is certain enough that the gift of ornament in natural things has not been lavished, as it is lavished, for the mere admiration of mankind. Ornament was as universal—applied upon a scale at once as grand and as minute as now—during the long ages before Man was born. Some of the most beautiful forms in Nature are the shells of the marine Mollusca, and many of them are the richest, too, in surface ornament. But, prodigal of beauty as the Ocean now is in the creatures which it holds, its wealth was even greater and more abounding in times when there was no man to gather them. The shells and corals of the old Silurian Sea were as elaborate and as richly carved as those which we now admire: and the noble Ammonites of the Secondary ages must have been glorious things indeed. Even now there is abundant evidence that although Man was intended to admire beauty, beauty was not intended only for Man's admiration. Nowhere is ornament more richly given, nowhere is it seen more separate from use, than in those organisms of whose countless millions the microscope alone enables a few men for a few moments to see a few examples. There is no better illustration of this than a class of forms belonging to the border-land of animal and vegetable life called the *Diatomaceæ*, which, though invisible to the naked eye, play an important part in the economy of Nature. They exist almost everywhere, and of their remains whole strata, and even mountains, are in great part composed. They have shells of pure silex, and these, each after its own kind, are all covered with the most elaborate ornament—striated, or fluted, or punctured, or dotted in patterns which are mere patterns, but patterns of perfect, and sometimes of most complex, beauty. No graving done with the graver's tool can equal that work in gracefulness of design, or in delicacy and strength of touch. Yet it is impossible to look at these forms—in all the variety which is often crowded under a single lens—without recognising instinctively that the work of the graver is work strictly analogous, —addressed to the same perceptions,—founded on the same idea,—having for its object the same end and aim. And as the work of the graver varies for the mere sake of varying, so does the work on these microscopic shells. In the same drop of moisture there may be some dozen or twenty forms, each with its own distinctive pattern, all as constant as they are distinctive, yet having all apparently the same habits, and without any perceptible difference of function.

It would be to doubt the evidence of our senses and of our reason, or else to assume hypotheses of which there is no proof whatever, if we were to doubt that mere ornament, mere variety, are as much an end and aim in the workshop of Nature as they are known to be in the workshop of the goldsmith and the jeweller. Why should they not? The love and desire of these is universal in the mind of Man. It is seen not more distinctly in the highest forms of civilized art than in the habits of the rudest savage, who covers with elaborate carving the handle of his war-club, or the prow of his canoe. Is it likely that this universal aim and purpose of the mind of Man should be wholly without relation to the aims and purposes of his Creator? He that formed the eye to see beauty, shall He not see it? He that gave the human hand its cunning to work for beauty, shall His hand never work for it? How, then, shall we account for all the beauty of the world—for the careful provision made for it where it is only the secondary object, not the first? Even in those cases, for example, where concealment is the main object in view, ornament is never forgotten, but lies as it were underneath, carried into effect under the conditions and limitations imposed by the higher law and the more special purpose. Thus the feathers of the Ptarmigan, though confined by the law of assimilative colouring to a mixture of black and white or gray, have those simple colours disposed in crescent bars and mottlings of beautiful form, even as the lichens which they imitate spread in radiating lines and semi-circular ripples over the weather-beaten stones. It is the same with all other Birds whose colour is the colour of their home. For the purpose of concealment, their colouring would be equally effective if it were laid on without order or regularity of form. But this is never done. The required tints are always disposed in patterns, each varying with the genus and the species; varying for the mere sake of variation, and for the beauty which belongs to ornament. And where this purpose is not under the restraint of any other purpose controlling it and keeping it down as it were within comparatively narrow limits, how gorgeous are the results attained! What shall we say of flowers—those banners of the vegetable world, which march in such various and splendid triumph before the coming of its fruits? What shall we say of the Humming Birds—whose feathers are made to return the light which falls upon them, as if rekindled from intenser fires, and coloured with more than all the colours of all the gems?

There is one instance in Nature (and, as far as I know, only one) in which ornament takes the form of pictorial representation. The secondary feathers in the wing of the Argus Pheasant are developed into long plumes, which the bird can erect and spread out like a fan, as a Peacock spreads his train. These feathers are decorated with a series of conspicuous spots or "eyes," which are so coloured as to imitate the effect of balls. The shadows and the "high light" are placed exactly where an artist would place them in order to represent a sphere.[1] The "eyes" of the Peacock's train are wonderful examples of ornament; but they do not represent anything except their own harmonies of colour. The "eyes" of the Argus Pheasant are like the "ball and socket" ornament which is common in the decorations of human art. It is no answer to this argument in respect to beauty, that we are constantly discovering the use of beautiful structures in which the beauty only, and not the usefulness, had been hitherto perceived. The harmonies on which all beauty probably depends are so minutely connected in Nature that "use" and ornament may often both arise out of the same conditions. Thus, some of the most beautiful lines on the surface of shells are simply the lines of their annual growth, which growth has followed definite curves, and it is the "law" of these curves that is beau-

[1] I owe the observation of this curious fact to my friend Mr. James Nasmyth, so well known as the inventor of the Steam Hammer, and as a distinguished astronomer.

tiful in our eyes. Again, the forms of many Fish which are so beautiful, are also forms founded on the lines of least resistance. The same observation applies to the form of the bodies and of the wings of Birds. Throughout Nature, ornament is perpetually the result of conditions and arrangements fitted to use and contrived for the discharge of function. But the same principle applies to human art, and few persons are probably aware how many of the mere ornaments of architecture are the traditional representation of parts which had their origin in essential structure. Yet who would argue from this fact that ornament is not a special aim in the works of Man? When the savage carves the handle of his war-club, the immediate purpose of his carving is to give his own hand a firmer hold. But any shapeless scratches would be enough for this. When he carves it in an elaborate pattern, he does so for the love of ornament, and to satisfy the sense of beauty.

There is, however, another department of natural phenomena which, much more than the one we have been now considering, does at first sight suggest to the mind the subordination of Purpose and the supremacy of Law. It is the department of Comparative Anatomy. It is a fact now well known and universally accepted, that in many animal structures, perhaps in all except one, there are parts the presence of which cannot be explained, from their serving any immediate use, or discharging any actual function. For example, the limbs of all the Mammalia, and even of all the Lizards, terminate in five jointed bones or fingers. But in many animals the whole five are not needed, but only some one, or two, or three. In such cases the remainder are indeed dwarfed, sometimes almost extinguished; but the curious fact is that rudimentally the whole number are always to be traced. Even in the Horse, where one only of the five is directly used, and where this one is enlarged and developed into a hoof, parts corresponding to the remaining four fingers can be detected in the anatomy of the limb. Other examples of the same principle might be given without number. Thus there are Monkeys which have no thumbs for use, but only thumb-bones hid beneath the skin: the wingless Bird of New Zealand, the "Apteryx," has useless wing-bones similarly placed: snakes destined always to creep "upon their belly" have nevertheless rudiments of legs, and the common "Slowworm" has even the "blade bone" and "collar bone" of rudimentary or aborted limbs: the Narwhal has only one tusk, on the left side, developed for use, like the horn of an heraldic Unicorn, but the other tusk, on the right side, is present as a useless germ: the female Narwhal has both tusks reduced to the same unserviceable condition: young whalebone Whales are born with teeth which never cut the gum, and which are afterwards absorbed as entirely useless to the creature's life.

At first sight it may appear as if these were facts not to be reconciled with the supremacy of Purpose :—at first sight, but at first sight only. For as we look at them and wonder at them, and set ourselves to discover how many of a like nature can be found, our eye catches sight of an Order which had not been at first perceived. Exceptions to one narrow rule such as we might have laid down and followed for ourselves, they are now seen to be in strict subordination to a larger rule which it would never have entered into our imagination to conceive. These useless members, these rudimentary or aborted limbs which puzzled us so much, are parts of an universal Plan. On this plan the bony skeletons of all living animals have been put together. The forces which have been combined for the moulding of Organic Forms have been so combined as to mould them after certain types or patterns. And when Comparative Anatomy has revealed this fact as affecting all the animals of the existing world, another branch of the same science comes in to conform the generalisation, and extend it over the innumerable creatures which have existed and have passed away. This one Plan of Organic Life has never been departed from since Time began.

When we have grasped this great fact, all the lesser facts which are subordinate to it assume a new significance. In the first place a Plan of this kind is in itself a Purpose. An Order so vast as this, including within itself such variety of detail, and maintained through such periods of Time, implies combination and adjustment founded upon, and carrying into effect, one vast conception. It is only as an Order of Thought that the doctrine of Animal Homologies is intelligible at all. It is a Mental Order, and can only be mentally perceived. For what do we mean when we say that this bone in one kind of animal corresponds to such another bone in another kind of animal? Corresponds—in what sense? Not in the method of using it—for very often limbs which are homologically the same are put to the most diverse and opposite uses. To what standard, then, are we referring when we say that such and such two limbs are homologically the same? It is to the standard of an Ideal Order—a Plan—a Type—a Pattern mentally conceived. This sounds very recondite and metaphysical; and yet the habit of referring physical facts to some ideal standard and order of thought is a universal instinct in the human mind. It is one of the earliest of our efforts in endeavouring to understand the phenomena around us. The science of Homologies, as developed by Cuvier and Hunter and

Owen and Huxley, is indeed an intricate, almost a transcendental science. Yet Dr. Livingstone found the natives of Africa debating a question which belongs essentially to that science and involves the whole principle of the mental process by which it is pursued. The debate was on the question "whether the two toes of the Ostrich represent the thumb and forefinger in Man, or the little and ring finger."[1] This is purely a question of Comparative Anatomy. It is founded on the instinctive perception that even between two frames so widely separated as those of an Ostrich and a Man, there is a common Plan of structure, with reference to which plan, parts wholly dissimilar in appearance and in use, can nevertheless be identified as "representative" of each other:—that is, as holding the same relative place in one Ideal Order of arrangement.

The recognition of this idea in minds so rude is not the less remarkable from the fact that both sides in this African debate were wrong in their practical application of the idea to the particular case before them. Unity of design amidst variety of form is so conspicuous and universal in the works of Nature that the perception of it could not possibly escape recognition even by the rudest human mind, formed as that Mind is to see

[1] "The Zambesi and its Tributaries," p. 424.

Order, and to work for it, and to admire it. But though instinct is enough to give us the general idea, and to trace it in a thousand instances where it can hardly be overlooked, yet it needs close and laborious study, and high powers of analysis and of thought, to trace correctly the true Order and Plan through the fine and subtle passages of Nature. It would have astonished those poor natives of Africa to be told, as is the truth, that if they wished to find in the Ostrich the parts corresponding to their own middle finger, or ring-finger, or any other finger, they must look, not to the toes of the Ostrich, but to her little aborted wings, which, though useless for the purposes of flight, are still retained as representing the wings of other Birds, and the forearms of all the Mammals.

For here we come upon the interchange and crossing as it were of two distinct ideas, which seem to stand the one as the warp and the other as the woof in the fabrics of Organic Life. There is the idea of Homology in Structure and the idea of Analogy in Use. The one represents the Unity of Design, the other represents Variety of Function. It might have been supposed that these could not easily be reconciled—that where great differences in use and application are essential, rigid adherence to one pattern of structure would be an impediment in the way. But it is not so. The same bones in different animals are made subservient to the widest possible diversity of function. The same limbs are converted into paddles, and wings, and legs, and arms. And so it is with every other part of the skeleton and every other organ of the body. Indeed it is difficult to say whether the law of unity in design, or the law of variety in adaptation, is pushed to the greatest length. There are some cases in which the adaptation of form to special function is carried so far that all appearance of common structure is entirely lost. It is very difficult, for example, to persuade persons ignorant of the principles of anatomy that the Whale and the Porpoise are not Fish, that they breathe with lungs as Man breathes, that they would be drowned if kept long under water, and that, as they suckle their young, they belong to the same great Class, Mammalia. Living in the same element as Fish, and feeding very much as fishes feed, a similar outward form has been given to them, because that form is the best adapted for progression through the water. But that form has been, so to speak, *put on* round the Mammalian skeleton, and covers all the organs proper to the Mammalian Class. Whales and Porpoises, notwithstanding their form, and their habitat, and their food, are as separate from Fishes as the Elephant, or the Hippopotamus, or the Giraffe.

And when we remember that the immense variety of Organic Forms in the existing world does not exhaust the adaptability of their Plan, but that the still vaster varieties of all the extinct creations have circled round the same central Types, it becomes evident that these Types have had from the first a Purpose which has been well and wonderfully fulfilled. As a matter of fact, we see that the original conception of the framework of Organic Life has included in itself provisions for applying the principle of adaptation in infinite degrees. Its last development is in Man. In his frame there is no aborted member. Every part is put to its highest use:—highest, that is, in reference to the supremacy of Mind.[1] There are stronger arms, there are swifter limbs, there are more powerful teeth, there are finer ears, there are sharper eyes. There are creatures which go where he cannot go, and can live where he would die. But all his members are co-ordinated with one power—the power of Thought. Through this he has the dominion over all other created things—whilst yet as regards the type and pattern of his frame he has not a single bone or joint or organ which he does not share with some one or other of the Beasts that perish. It is not in any of

[1] "Quid reliquâ descriptione omnium corporis partium, in quâ nihil inane, nihil sine causâ, nihil supervacaneum est?"—Cicero, "De Nat. Deor.," lib. i. cap. 33.

the parts of his structure, but in their combination and adjustment, that he stands alone.

All these facts must convince us that we must enlarge our ideas as to what is meant by Use in the Economy of Nature. In the first place, it must be so interpreted as to include ornament; and in the second place, it must include also not merely Actual Use, but Potential Use, or the capacity of being turned to use in new creations. Of course this is one of the ideas which Philosophers of the Positive School denounce as "Metaphysical." But here again their opposition is itself based upon metaphysics, only upon metaphysics which are bad. "Potential existence," says Mr. Lewes,[1] "is ideal not real." "A fact is not a fact until it is accomplished. Nothing exists before it exists. This truism is disregarded by those who talk of potential existence." So it is, and it ought to be disregarded, because it has no bearing on the question. May not the formation of a plan or conspiracy to murder be "a fact" although the murder is not "accomplished?" Is not the capacity in the different pieces of a puzzle of being fitted together, a fact—even when the pieces are all huddled confusedly in a box? Is there no potential use in the udder of a cow-calf before it can have had any calves of its own? Is the idea of Potential

1 "History of Philosophy," Prologue, p lxxxviii.

use in all these cases an idea which has no "reality?" Are they mere "artifices of thought," or "preliminary falsifications of fact?" If the metaphysics of Positivism are available to establish this conclusion, they must be equally available to condemn knowledge in all its forms as "Ideal" and not "real." Bad metaphysics of this kind are indeed, what Dr. Newman dreads the human mind may be, a "universal solvent," casting doubt on the most certain of its own conclusions, and landing itself in universal scepticism.

We have not far to go to find the same kind of reasoning, and the same methods of analysis, employed to establish the converse proposition, that so far from Potentiality having no existence, it is the only form under which the existence of anything beyond ourselves can be known to us. No less eminent a thinker than Mr. J. S. Mill reduces Matter itself, and the very idea of the existence of an external world, to a "Permanent Possibility of Sensation."[1] Nay, he is not sure—he only sees some "intrinsic difficulties" in the way—whether our knowledge of Self-existence may not be brought under the same "Potential" category—as a mere "Possibility of Sensation."[2] In regard to Matter, Mr. Mill distinctly says that so far from a mere Possibility having no real

[1] "Mill on Hamilton," chap. xi. [2] Ibid. chap. xii.

existence, it is the only reality—the one thing which is constant and abiding behind the flux and uncertainty of actual sensations. My own opinion is that the metaphysical process by which these opposite paradoxes are arrived at is nearly as worthless in the one case as in the other. Of the two I prefer the paradox of Mr. Mill to the paradox of Mr. Lewes—so far at least as the reality of Potential Existences is concerned. But I prefer it only in the very case to which Mr. Mill shrinks from applying it. I *can* think of my own mind or existence as a "Possibility of Sensation" (whether "permanent" or not). It is a method of conception indeed which casts no light on anything, and it is highly artificial; but at least it is not false. It involves no confounding of two different elements of thought. But I cannot transfer the word or the idea of sensation from myself to the external things which cause sensation in me. This transfer involves a fundamental confusion of thought and of language as the instrument of thought.[1] But such paradoxes are the natural result of one great error —the endeavour to get rid of, or to explain away, or to dissolve by analysis, such simple and elementary conceptions of the Mind as the idea of External Force and of Causation, or the idea of Purpose and Intention.

[1] See Note D.

Matter may very well be conceived as "That which produces, or has a Possibility of producing, Sensation in Sentient beings." But this is a definition which involves the idea of Causation. And if this be rejected as an elementary conception, (or as a distinct conception, whether elementary or not,) then the paradox of Mr. Mill is the natural result. In like manner, if the idea of Purpose and Intention be repudiated, as representing no "reality" in Nature, then the opposite paradox of Mr. Lewes is reached along the same slippery and deceptive ways. We know at least, as a matter of experience, that we are capable of forming plans which exist as such before they are carried into effect. We know too that one plan may be large enough to include another, and that even within the fractional limits of our foresight we can provide for contingent as well as for actual use. We can therefore easily conceive the existence of the same kind of prevision in the Mind which works in Nature, and we can easily understand how the apparent difference between actual and contingent use should be greater in proportion as the Plan is larger, and is designed to operate during vaster periods of Time.

In this point of view rudimentary or aborted organs need no longer puzzle us, for in respect to Purpose they may be read either in the light of History, or in the light of Prophecy. They may be regarded as indicating always either what had already been, or what was yet to be. Why new creations should never have been made wholly new;—why they should have been always moulded on some pre-existing Forms;—why one fundamental ground-plan should have been adhered to for all Vertebrate Animals, we cannot understand. But as a matter of fact it is so. For it appears that Creative Purpose has been effected through the instrumentality of Forces so combined as to arrange the particles of organic matter in definite forms: which forms include many separate parts having a constant relation to each other and to the whole, but capable of arrestment or development according as special organs are required for the discharge of special functions. Each new creation seems to have been a new application of these old materials. Each new House of Life has been built on these old foundations. Among the many wonders of Nature there is nothing more wonderful than this—the adaptability of the one Vertebrate Type to the infinite variety of Life to which it serves as an organ and a home. Its *basement* has been so laid that every possible change or addition of superstructure could be built upon it. Creatures destined to live on the earth or in the earth, on the sea or in the sea, under every variety of condition of existence, have all been made after that one pattern; and each of them with as close an adaptation to special function as if the pattern had been designed for itself alone. It is true that there are particular parts of it which are of no use to particular animals. But there is no part of it which is not of indispensable use to some member of the group; and there is one Supreme Form in which all its elements receive their highest interpretation and fulfilment. It is indeed wonderful to think that the feeble and sprawling paddles of a Newt, the ungainly flippers of a Seal, and the long leathery wings of a Bat, have all the same elements, bone for bone, with that human hand which is the supple instrument of Man's contrivance, and is alive, even to the finger-tips, with the power of expressing his Intellect and his Will. Here again the Laws of Nature are seen to be nothing but combinations of Force with a view to Purpose: combinations which indicate complete knowledge, not only of what is, but of what is to be, and which foresees the End from the Beginning.

CHAPTER V.

CREATION BY LAW.

WE see, then, how the existence of Organs separated from Function, and of structures without immediate use, find their natural place among all the other phenomena of the world. They do not show that "Law" is ever superior to Will, or can ever assert, even for a moment, an independence of its own. On the contrary, they show, as nothing else can show, the patient movements, and the incalculable years, through which material laws have been made to follow the steps of Purpose.

But, then, let us remember this: these discoveries in Physiology, though they are helpless to prove that Law has ever been present as a Master, are eminently suggestive of the idea that Law has never been absent as a Servant;—that as, in governing the world, so in forming it, Material Forces have been always used as the instruments of Will.

It is no mere theory, but a fact as certain as any other fact of Science, that Creation has had a History. It has not been a single act, done and finished once for all, but a long series of acts—a work continuously pursued through an inconceivable lapse of time. It is another fact, equally certain, respecting this work, that as it has been pursued in Time, so also it has been pursued by Method. There is an "observed Order of facts" in the history of Creation, both in the organic and in the inorganic world. I speak here, however, of the organic world alone, and chiefly of those higher Forms which are the seat of Animal Life. In these, there is an observed Order in the most rigid scientific sense—that is, phenomena in uniform connexion, and mutual relations which can be made, and are made, the basis of systematic classification. These classifications are imperfect, not because they are founded on ideal connexions where none exist, but only because they fail in representing adequately the subtle and pervading Order which binds together all living things. But the Order which prevails in the existing world is not the only Order which has been recognised by science. A like Order has prevailed through all the past history of Creation. Nay, more; it has, I think, been clearly ascertained, not only that relations similar to those which now exist have existed always among all the animals of each contemporary Creation, but that Order of a like kind has connected with each other all the different Creations which were successively introduced. In almost all the leading

Types of Life which have existed in the different geological ages, there is an orderly gradation connecting the Forms which were becoming extinct with the Forms which were for the first time appearing in the world. It is still disputed by some geologists, whether we have certain evidence that this gradation has been the gradation of a rising scale of progressive Creations from lower to higher Types. But this dispute is maintained only on the ground that we cannot safely trust to negative evidence. It is an unquestionable fact, that so far as this kind of evidence can go, it does testify to the successive introduction of higher and higher Forms of Life. Very recently a discovery has been made, to which Mr. Darwin only a few years ago referred as "a discovery of which the chance is very small"—viz. of fossil Organisms in beds far beneath the lowest Silurian strata. This discovery has been made in Canada—in beds far down, near the bottom even, of the rocks hitherto termed "Azoic." But what are the Forms of Life which have been found here? They belong to the very lowest of living types—to the "Rhizopods." So far as this discovery goes, therefore, it is in strict accordance with all the facts previously known—that, as we go back in time, we lose, one after another, the higher and more complex organisms: first, the Mammalia; then the Vertebrata; and now, lastly, even the Mollusca. It is in accordance, too, with another fact which has been observed before, viz. that particular Forms of Life have attained, at particular epochs, a maximum development, both in respect to size and distribution—the favourites, as it were, of Creation for a time. These earliest Rhizopods seem to have been of enormous size, and developed on an enormous scale; since there is good reason to believe that beds of immense thickness are composed of their remains. All that is new in this discovery is the vast extension which it gives in Time to the same rules which had been already traced through ages which we cannot number.

Then, there is another observed Order. For each Class of animal some definite Type or pattern has been adhered to; and the modifications of that Type have been gradual and successive. In many cases the science of fossil remains enables us to trace the intermediate Forms through which existing animals can be connected with animals long since extinct. It must be remembered that the fact of this connexion is quite a separate thing from any theory as to its physical cause. Professor Owen pointed out some years before the publication of Mr. Darwin's theory, the existence of fossil animals which showed an increasing approximation to the forms of the Horse and of the Ox: and he showed that this approximation was related in Time, as it seemed to be in Purpose, with the coming need of them for the service and use of Man. These are the facts on which the idea of "Creation by Law" is founded. Let us look a little nearer what this idea is, and what it involves. It is an idea much vaunted by some men, much feared by others. Perhaps it may be found, on closer investigation, that they are fearing or worshipping, as the case may be, an idol of the imagination.

It being certain that Creation exhibits an Order of facts which can be so clearly defined and traced, it follows, that at least in this first sense of the word, Creation has been by Law. We are, therefore, led on to the farther question, whether Law in any other sense can be traced or detected in the work of Creation? Is the observed Order which prevails in the organic world an Order of which we can even guess the physical cause? Is it an Order which contains within itself any indications of the Force or combination of Forces which have been concerned in producing it?

In considering this question, there is one thing to be observed at the outset. It is certain that nothing is known, or has been even guessed at, in respect to the history and Origin of Life, which corresponds with Law in its strictest and most definite sense. We have no knowledge of any one or more Forces—such as the Force of Gravitation, or of magnetic attraction and repulsion—to which any one of the phenomena of Life can be traced. Far less have we any knowledge of any laws of the like kind which can be connected with the successive creation or development of new Organisms. Professor Huxley, in a recent work,[1] has indeed spoken of "that combination of natural forces which we term Life." But this language is purely rhetorical. I do not mean to say that Life may not be defined to be a kind of Force, or a combination of Forces. All I mean is, that we know nothing of any of these Forces in the same sense in which we do know something of the Force of Gravity, or of Magnetism, or of Electricity, or of Chemical Affinity. These are all more or less known, not, indeed, in respect to their ultimate nature, but in respect to certain methods and measures of their operation. No such knowledge exists in respect to any of the Forces which have been concerned in the development of Life. No man has ever pretended to get such a view of any of these as to enable him to apply to them the instruments of his analysis, or to trace in their working any definite relations to Space, or Time, or Number.

Since, then, laws, in this most definite sense of the word, have not been discovered in the existing pheno-

[1] "Elements of Comparative Anatomy," p. 2.

mena, or in the past history of Organic Life, let us look a little closer at the ideas which these phenomena have suggested to the mind of those who have speculated on the Origin and Development of Species.

There is one idea which has been common to all theories of Development, and that is, the idea that ordinary generation has somehow been producing, from time to time, extraordinary effects, and that a new Species is, in fact, simply an unusual birth. It is worthy of observation, that the earlier forms in which the theory of Development appeared, did suggest something more nearly approaching to a Law of Creation than is contained in the later form which that theory has assumed in the hands of Mr. Darwin. The essential idea of the theory of Development, in its earlier forms, was, that modifications of structure arose somehow by way of natural consequence from the outward circumstances or physical conditions which required them, and from the living effort of Organism sensible in some degree of that requirement. Now, inadequate and even grotesque though this idea may be as explaining the Origin of new Species, it cannot be denied, that it makes its appeal to a process which, at least to a limited extent, does operate in producing modifications of organic structure. For example, the same species of Mollusc has often a shell comparatively weak and thin, or a shell comparatively robust and strong, according as it lies in tranquil or in stormy water. The shell which is much exposed needs to be stronger than the shell which is less exposed. But it is obvious that the mere fact of the need cannot supply the thing needed, unless by the adjustment of some machinery for the purpose. How the vital forces of the Mollusc can thus be made to work to order, under a change of external conditions, we do not know. But we do know, as a matter of fact, that the shell is thickened and strengthened, according as it needs resisting power. This result does not appear to arise from any difference in the amount of lime held in solution in the water, but from some power in the secreting organs of the animal to appropriate more or less of it, according to its own need. The effects of this power are seen where there is no difference of condition except difference of exposure. It is said that they are observable, for example, in the shells which lie on the different sides of Plymouth Breakwater,—the sheltered side and the exposed side. The same power of adaptation is seen in many other forms. Trees which are most exposed to the blast are the most strongly anchored in the soil. Limbs which are the most used are the most developed. Organs which are in constant use, are strengthened, whilst organs in habitual disuse have a tendency to become weaker.

All these results arise by way of natural consequence. How shall we describe them? Shall we say that they are the result of Law? We may safely do so, remembering only that by Law, in this sense, we mean nothing but the co-operation of different natural Forces, which, under certain conditions, work together for the fulfilment of an obvious intention. Of the nature of those Forces we know nothing; nor is it easy to conceive how they have been so co-ordinated as to produce effects fitting with such exactness into the conditions requisite for the preservation of Organic Life. If there were any evidence that by the same means new Forms of Life could be developed from the old, I cannot see why there should be any reluctance to admit the fact. It would be different from anything that we see; but I do not know that it would be at all less wonderful, or that it would bring us much nearer than we now stand to the great mystery of Creation. The adaptation and arrangement of natural forces, which can compass these modifications of animal structure, in exact proportion to the need of them, is an adaptation and arrangement which is in the nature of Creation. It can only be due to the working of a power which is in the nature of Creative power.

We are so accustomed to these and other similar phenomena, and so apt to hide our own ignorance of their cause, by describing them as the result of "Law," that we forget what a multitude of natural Forces must be concerned in their production, and what complicated adjustments of these amongst each other for the accomplishment of Purpose. It is purely, therefore, in my view, a question of evidence, whether this particular law of adaptation has or has not been the means of introducing new Forms of Life. There is no evidence that it has. So far as we know, this power of self-adaptation, wonderful as it is, has a comparatively limited application; when that limit is outrun by changes in outward conditions, which are too great or too rapid, whole Species die and disappear. Nevertheless, the introduction of new Species to take the place of those which have passed away, is a work which has been not only so often, but so continuously repeated, that it does suggest the idea of having been brought about through the instrumentality of some natural process. But we may say with confidence, that it must have been a process different from any that we yet know—a process not the same as that (obscure as that is) which produces the lesser modifications of Organic Forms.

It has not, I think, been sufficiently observed, that the theory of Mr. Darwin does not address itself to the same question, and does not even profess to trace the Origin of new Forms to any definite law. His theory gives an explanation, not of the processes by which new Forms first appear, but only of the processes by which, when they have appeared, they acquire a preference over others, and thus become established in the world. A new Species is, indeed, according to his theory, as well as with the older theories of Development, simply an unusual birth. The bond of connexion between allied specific and generic Forms, is in his view simply the bond of Inheritance. But Mr. Darwin does not pretend to have discovered any law or rule according to which new Forms have been born from old Forms. He does not hold that outward conditions, however changed, are sufficient to account for them. Still less does he connect them with the effort or aspirations of any Organism after new faculties and powers. He frankly confesses that "our ignorance of the laws of variation is profound;" and says, that in speaking of them as due to chance, he means only "to acknowledge plainly our ignorance of the cause of each particular variation."[1] Again he says—"I believe in no law of necessary development."[2]

This distinction between Mr. Darwin's theory and other theories of Development, has not, I think, been sufficiently observed. His theory seems to be far better

[1] "Origin of Species," p. 131 (first edition). [2] Ibid. p. 351.

than a mere theory—to be an established scientific truth—in so far as it accounts, in part at least, for the success and establishment and spread of new Forms *when they have arisen.* But it does not even suggest the law under which, or by which, or according to which, such new Forms are introduced. Natural Selection can do nothing except with the materials presented to its hands. It cannot select except among the things open to selection. Natural Selection can originate nothing; it can only pick out and choose among the things which are originated by some other law. Strictly speaking, therefore, Mr. Darwin's theory is not a theory on the Origin of Species at all, but only a theory on the causes which lead to the relative success or failure of such new Forms as may be born into the world. It is the more important to remember this distinction, because it seems to me that Mr. Darwin himself frequently forgets it. Not only does he speak of Natural Selection "producing" this and that modification of structure, but he undertakes to affirm of one class of changes that they can be produced, and of another class of changes that they cannot be produced by this process.[1]

Now, what are the changes for the preservation of

[1] "Origin of Species," p. 200 (first edition).

which Natural Selection does, in some sense, account? They are such changes, and these only, as are of some direct use to the Organism in the "struggle for existence." Any change which has not this direct value, is not provided for in the theory. All structures, therefore, are unaccounted for—not only as respects their origin, but even as respects their preservation—in which the variations have no other value than mere beauty or variety. Accordingly, Mr. Darwin is tempted, as I have already had occasion to observe, to deny that any such structures xist in Nature. Any theory of which this denial is really a necessary part, is self-condemned. Yet a theory may be good as accounting for the preservation of some structures, although it fails to account for the preservation of others. And so the fact that Natural Selection cannot have operated on structures of mere beauty and variety is no proof that the theory of Natural Selection is false, but only that it is incomplete. It does not account for the origin of any structure; and it accounts for the preservation of only a certain number. Surely, then, Mr. Darwin assigns to his "law" of Natural Selection a range far wider than really belongs to it, when, on the strength of it, he denies that beauty for its own sake can be an end or object in Organic Forms. He says—"This doctrine, if true, would be absolutely fatal to my theory." Why should this be fatal to his theory, except on the supposition that Natural Selection gives a *complete* account both of the Origin of new Forms, (of which, in reality, it gives no account at all,) and of their preservation, of which it does give some account, but one which is only partial? I dwell on this, because it lies at the very root of the question, how far Mr. Darwin's theory can be said to suggest anything in the nature of a Creative Law of a kind to explain the Method which has been followed in the introduction of new Forms.

We may test this question by bringing to bear upon it some particular example of specific variation. I select for this purpose one example, which will illustrate the subject better than any abstract discussion. It is the case of the Humming Birds.

This group of Birds seems to exhibit, in the most striking form, not a few of those mysteries of Creation which at once tempt us to speculate on the Origin of Species, and at the same time confound every endeavour to bring it into relation with any process which we know or can conceive. In the first place, they are sharply defined from all other forms in that Class of the animal kingdom to which they belong. It is most difficult to say what is their nearest affinity, and the nearest, when it is found, is very distant. Secondly, they are absolutely confined to one Continent of the Globe. In the third place, the various Species as amongst themselves are very closely united, ranging, indeed, over a great variety of forms, but for the most part connected with each other by very nice gradations. In the fourth place, there are, so to speak, some gaps in the scale, which suggest that some Species have either been lost, or have not yet been discovered. In the fifth place, each of these Species, however nearly allied to some other, appears to be absolutely fixed and constant, there being not the slightest indication of any mixture—of any hybrid forms. In the sixth place, there is the most wonderful adaptation of special organs for the performance of special functions, and for the relation of these organs to particular structures in the vegetable kingdom. In the seventh place, there is a development, for which, in extent and variety, there is no parallel in the world, of structures designed apparently for mere ornament, and entirely separate from any other known or conceivable use.

A few words on some of these characters will show their separate and joint bearing on the idea of Creation by Law.

In the first place, then, the absolute distinctiveness from all others of this Family of Birds, coupled with its immense extent, gives the idea of some common bond, some physical cause, to which such an identity in physical characters must be due. This identity prevails not only in such essential matters as the structure of the bill and tongue, in the form of the feet and of the wings, in the habits of flight, and in the nature of the food, but runs also into some very curious details, as, for example, in the number of feathers in the tail and in the wings, which are constant numbers—adhered to even when some of the feathers, not being used even for ornament, are reduced almost to rudiments. But under degrees of development which are very variable, the number is invariable. This identity of structure is the more remarkable from the immense extent of the group which it characterises. There are now known to science no less than about 430 different species of Humming Bird; and it cannot be doubted that many more remain to be discovered among the immense forests and mountain ranges of Central America.

Now, what is the bond that unites so closely, in a common structure, all the forms of this great Family of birds? We think it a sufficient explanation sometimes of the likeness of things, that they are made for a common purpose. And so it is an explanation in one sense, but not in another. It gives the reason why likeness should be aimed at, but not the cause or the means through which it has been brought about. Sameness in the purpose for which things are intended, is a reason why those things should be made alike; but it is no explanation of the process to which the common aspect is due. It is an explanation of the "why;" but it is no explanation of the "how." Purpose is attained in Nature through the instrumentality of means; and community of aspect in created things suggests the idea of some common process in the creative work. Thus, the likeness which is due to common parentage serves the most important purposes; but it is not the less the result of a physical cause, out of which it arises by way of natural consequence. The likeness of the Humming Birds to each other suggests this kind of cause. It is true that the organs which it principally affects are specially adapted for a special habit of life. They are fitted to enable the Bird to feed on the nectar, and the insects which frequent the nectar of flowers, or the leaves or bark of trees. But there are flowers and insects in abundance in other quarters of the globe where there are no Humming Birds.

And here we come on the curious facts of geographical distribution,—a class of facts which, as much as any other, suggest some specific methods as having been followed in the work of Creation. Humming Birds are absolutely confined to the great Continent of America with its adjacent islands. Within those limits there is every range of climate, and there are particular species of

Humming Bird adapted to every region where a flowering vegetation can subsist. It is therefore neither climate nor food which confines the Humming Birds to the New World. What is it, then? The idea of "centres of Creation" is at once suggested to the mind. It seems as if the Humming Birds were introduced at one spot, and as if they had spread over the whole Continent which was accessible to them from that spot. They are absent elsewhere, simply because from that spot the other Continents of the world were inaccessible to them. But if these ideas are suggested to the mind by the general aspect of this family as a whole, they are strengthened by some of the facts which we discover when we examine and compare with each other the genera and species of which it is composed. There is a beautiful gradation between the different genera and the different species,—so much so, that it has been found impossible to divide the Humming Birds into more than two sub-families, from the absence of sufficiently well-marked divisions. And yet on the other hand, they cannot be arranged in anything like a continuous series, because some links appear to be missing in the chain.

But these general facts terminate in nothing more definite than a vague surmise. When we enter farther into details, we feel at once how little they agree with any physical law which is known or even conceivable by us. If the likeness which prevails in the whole group reminds us of the likeness which is due to community of blood, it is equally true that the differences between the species are totally distinct, both in kind and degree, from the variation which we ever see arising among the offspring of the same parents. Let us look at what these differences are. The generic and specific distinctions between the Humming Birds are mainly of two kinds,—1*st*, Differences in the form of essential organs, such as the bill and the wings; 2*d*, Differences in those parts of the plumage which are purely ornamental. Now, of these two kinds of variation, the only one on which the law of Natural Selection has any bearing at all is the first. And on that kind of variation, the only bearing which Natural Selection has is this—that if any Humming Bird were born with a new form of bill, or a new form of wing, which enabled it to feed better and to range farther, then that improved bill and wing would naturally tend to be perpetuated by ordinary generation. This is unquestionably true; but it really does not touch the facts of the case. The bills and wings of the different genera do not differ from each other in respect of any comparative advantage of this kind, but simply in respect to variety corresponding with the variety of certain vegetable Forms.

One form of bill is as good as another, but some forms are adapted to some special class of flower. Some bills, for example, are formed of enormous length, specially adapted to obtain access to the nectar chambers of long tubular flowers, such as the *Brugmansia*. Some, on the other hand, as if to show that the same end may be attained by different means, obtain access to the same flowers by a shorter process, and pierce the bases of the corolla instead of seeking access by the mouth. Some have bills bent downwards like a sickle, adapted to searching the bark of Palm-trees for the insects hid under the scaly covering; others have bills curved in the opposite direction, fitted, apparently, to the curious construction of some of the great family of Orchids so immensely developed in the forests of Central America. Some have bills equally well adapted for searching a vast variety of flowers and blossoms, and these, accordingly, migrate with the flowering season, and, issuing from the great stronghold of the family in tropical America, spread like our own summer Birds of passage, northwards to Canada, and southwards to Cape Horn, in the corresponding seasons of the year. In contrast with these species of extended range, there are many species whose habitat is confined, perhaps to a single mountain, and there are a few which never have been seen beyond the edges of some extinct volcano, whose crater is now filled with a special flora. Many of the great mountains of the Andes have each of them species peculiar to themselves. On Chimborazo and Cotopaxi, and other summits, special forms of Humming Birds are found in special zones of vegetation even close up to the limits of perpetual snow. Again, many of the Islands have species peculiar to themselves. The little island of Juan Fernandez, 300 miles from the mainland, has three species peculiar to itself, of which two are so distinct from all others known, that they cannot for a moment be confounded with any of them.[1]

It is impossible not to see, in such complicated facts as these, that the creation of new Species has followed some plan in which mere variety has been in itself an object and an aim. The divergence of form is not a divergence which can have arisen by way of natural consequence, merely from comparative advantage and disadvantage in the struggle for existence. Bills highly specialised in form are certainly not those which would give the greatest advantage to birds which have equal access to the abundant Flora of an immense Continent. Some form of bill adapted to the probing or piercing of all flowers with almost equal ease, would seem to be the form most favourable to the multiplication and spread of

[1] Gould's "Trochilidæ."

Humming Birds. Continued approximation to some common type would seem to be quite as natural a change, and a much more advantageous kind of change as regards advantage in the struggle for existence, than endless divergence and special adaptation to limited spheres of enjoyment. At all events, we may safely say that mere advantage, in Mr. Darwin's sense, is not the rule which has chiefly guided Creative Power in the Origin of these new Species. It seems rather to have been a rule having for its object the mere multiplying of Life, and the fitting of new Forms for new spheres of enjoyment, according as these might arise out of corresponding changes in other departments of the organic world.

If, now, we turn to the other kind of specific distinction between Humming Birds, viz., that which consists in differences in the mere colouring and disposition of the plumage, we shall find the same phenomena still more remarkable. In the first place, it is to be observed of the whole group that there is no connexion which can be traced or conceived between the splendour of the Humming Birds and any function essential to their life. If there were any such connexion, that splendour could not be confined, as it almost exclusively is, to one sex. The female Birds are of course not placed at any disadvantage in the struggle for existence by their more sombre colouring. Mere utility in this sense, therefore, can have had no share in determining one of the most remarkable of all the characteristics of this family of Birds. It is obviously beside the question to account, as Mr. Wallace and Mr. Darwin do, for the beauty of the Humming Birds upon the ground that the males are thus rendered more attractive to the females. This attractiveness can only operate as between different individuals of the same species, since no one ever heard of the females of a dull-coloured species wandering in their affection from their rightful lords to the more brilliant males of some other species. Every animal, however little beautiful it may be in our eyes, has sufficient attractiveness as between the sexes to secure the great object of the continuation of its race. Utility, indeed, in a different sense, can be quoted with probability, as accounting for the comparative plain colouring of females in this and in almost all other genera of Birds. But then it is Utility conceived as operating by way of motive in a Creative Mind, and not operating as a physical cause in the production of a mechanical result. And here we find Mr. Wallace instinctively testifying to this great distinction, and employing language which indicates the passage from one order of ideas to another. He says, "The REASON WHY female birds are not adorned with equally brilliant plumes is sufficiently clear; they would be injurious by rendering their possessors too conspicuous during incubation."[1] This is, no doubt, the true explanation of the purpose which the plain colouring of female Birds is intended to serve; but it is no explanation at all of the physical causes by which this special protection is secured.

Those who, by special study, have laid their minds alongside the Mind of Nature in any of her Provinces, have generally imparted to them a true sense, so far as it goes, in the interpretation of her mysteries. Let us then hear what Mr. Gould says on the beauty of the Humming Birds:—"The members of most of the genera have certain parts of their plumage fantastically decorated; and in many instances most resplendent in colour. My own opinion is, that this gorgeous colouring of the Humming Birds has been given for the mere purpose of ornament, and for no other purpose of special adaptation in their mode of life; in other words, that ornament and beauty, merely as such, was the end proposed."[2] Different parts of the plumage have been selected in different genera as the principal subject of ornament. In some, it is the feathers of the crown worked into different forms of crest; in some, it is the feathers of the throat, forming gorgets and beards of many shapes and hues; in some, it is a special development of neck plumes, elongated

[1] *Quarterly Journal of Science*, Oct. 1867, p. 481.

[2] Gould's "Trochilidæ," Introduction.

into frills and tippets of extraordinary form and beauty. In a great number of genera the feathers of the tail are the special subjects of decoration, and this on every variety of plan and principle of ornament. In some, the two central feathers are most elongated, the others decreasing in length on either side, so as to give the whole the wedge form. In others, the converse plan is pursued, the two lateral feathers being most developed, so that the whole is forked after the manner of the common Swallow. In others, again, they are radiated or pointed and sharpened like thorns. In some genera there is an extraordinary development of one or two feathers into plumes of enormous length, with flat or spatulose terminations. Mere ornament and variety of form, and these for their own sake, is the only principle or rule with reference to which Creative Power seems to have worked in these wonderful and beautiful Birds. And if we cannot account for the differences in the general style and plan of ornament followed in the whole group, by referring them to any sort of use in the struggle for existence, still less is it possible to account, on this principle, for the kind of difference which separates from each other the different species in each of the genera. These differences are often little more than a mere difference of colour. The radiance of the ruby or topaz in one species, is replaced perhaps by the radiance of the emerald or the sapphire in another.

In all other respects the different species are sometimes almost exact counterparts of each other. As an example, let me refer to the two species figured by Mr. Gould as the Blue-tailed and the Green-tailed Sylphs; and also to two species of the "Comets," in which two different kinds of luminous reds or crimsons are nearly all that serve to distinguish the species.

A similar principle of variation applies in other genera, where the amount of difference is greater. For example, one of the most singular and beautiful of all the tribe is comprised within the genus *Lophornis*, or the "Coquettes." The principle of ornament in this genus is, that the different species are all provided both with brilliant crests, and with frills or tippets on the neck. The feathers of these parts are generally of one colour, ending in spots or spangles of another; the spangles being generally of metallic lustre. There seems to be a rule of inverse proportion between the two kinds of ornament. The species which have the neck plumes longest have the shortest crests, and *vice versâ*. In the shape and structure of all essential organs there is hardly any difference between the species.

One very curious example of variety for the sake of ornament may be mentioned in connexion with this wonderful family of Birds. It is a law—in the sense of an observed order of facts—regulating the ornament of Humming Birds, that where white is introduced into the colouring of the tail feathers, it is not applied to the central feathers, but is confined to the marginal feathers on either side. There is, however, one species (*Urosticte Bengamini*), recently discovered, which affords the only example yet known of a departure from this rule. It is a species in which white is one of the principal ornaments of the Bird, and is used in places where it can be placed in conspicuous contrast with the darkest tints. Tufts and lines of purest white shine among the greens and violets of the neck and head; whilst, in exquisite harmony with this, the four central feathers of the tail are alone dipped, as it were, in a solid glaze of the same white, and the marginal feathers on either side are kept wholly dark. Then, as if to mark with emphasis the meaning of this departure from the ordinary rule, it is a departure confined to the ornamented sex; and the Female Form of the same species follows the ordinary law—white being introduced in the marginal feathers, and in these alone.

Now, what explanation does the law of Natural Selection give—I will not say of the origin, but even of the continuance and preservation — of such specific varieties as these? None whatever. A crest of topaz is no better in the struggle for existence than a crest of sapphire. A frill ending in spangles of the emerald is no better in the battle of life than a frill ending in the spangles of the ruby. A tail is not affected for the purposes of flight, whether its marginal or its central feathers are decorated with white. It is impossible to bring such varieties into relation with any physical law known to us. It has relation, however, to a Purpose, which stands in close analogy with our own knowledge of Purpose in the works of Man. Mere beauty and mere variety, for their own sake, are objects which we ourselves seek when we can make the Forces of Nature subordinate to the attainment of them. There seems to be no conceivable reason why we should doubt or question, that these are ends and aims also in the Forms given to living Organisms, when the facts correspond with this view, and with no other. In this sense, we can trace a creative law,—that is, we can see that these Forms of Life do fulfil a purpose and intention, which we can appreciate and understand.

But then it may be asked, has this purpose and intention been attained without the use of means? Have no physical laws been used, whereby these new forms of beauty have been evolved, the one from the other, in a series so wonderful for its variety in unity, and its unity in variety? I am not now seeking to answer this question in the negative. All I say is, that the physical laws which are made subservient to this purpose are entirely unknown to us. That particular combination of a great many natural laws, which Mr. Darwin groups under the name of Natural Selection, does not in the least answer the conditions which we seek in a law to account for either the origin or the spread of such creatures as the various kinds of Humming Birds. On the other hand, if I am asked whether I believe that every separate Species has been a separate creation—not born, but separately made—I must answer, that I do not believe it. I think the facts do suggest to the mind the idea of the working of some creative Law, almost as certainly as they convince us that we know nothing of its nature, or of the conditions under which it does its glorious work. Our experience of the existing Order of Nature is, that the young of each species repeat the form and the colours of their parent, and that even where variations occur, they are inconstant, and tend to disappear. We have no knowledge, for example, that from the eggs of the Blue-tailed Sylph a pair of Green-tailed Sylphs can ever be produced. We have no reason to believe that a species of *Lophornis* with a tippet of emerald spangles, can ever hatch out a pair of young adorned with spangles of some other gem. And yet we cannot assert that such phenomena are impossible, nor can it be denied that, as a matter of speculation, this process is natural and easy of conception, as compared with the idea of each Species being separately called into existence, out of the inorganic elements of which its body is composed.

Such new births—if they do take place—would perfectly fulfil, I think, the only idea we can ever form of new creations. For example, it would appear that every variety which is to take its place as a new Species must be born male and female; because it is one of the facts of specific variation in the Humming Birds, that although the male and female plumage is generally entirely different, yet the female of each Species is as distinct from the female of every other, as the male is from the male of every other. If, therefore, each new variety were not born in couples, and if the divergence of Form were not thus secured in the organisation of both the sexes, it would fail to be established, or would exhibit for a time the phenomena of mixture, and terminate in reversion to the original type. Now here again we have the emphatic declaration of Mr. Gould, that among the thousands of specimens which have passed through his hands, from all the genera of this great family, he has never seen one case of mixture or hybridism between any two Species, however nearly allied. But this passage is so important, that I quote it entire. "It might be thought by some persons that four hundred species of birds so diminutive in size, and of one family, could scarcely be distinguished from each other; but any one who studies the subject, will soon perceive that such is not the case. Even the females, which assimilate more closely to each other than the males, can be separated with perfect certainty; nay, even a tail-feather will be sufficient for a person well versed in the subject to say to what genus and species the Bird from which it has been taken belongs. I mention this fact to show that what we designate a Species has really distinctive and constant characters; and in the whole of my experience, with many thousands of Humming Birds passing through my hands, I have never observed an instance of any variation which would lead me to suppose that it was the result of a union of two species. I write this without bias, one way or the other, as to the question of the Origin of Species. I am desirous of representing Nature in her wonderful ways as she presents herself to my attention at the close of my work, after a period of twelve years of incessant labour, and not less than twenty years of interesting study."[1]

If, therefore, new Species are born from the old, it is not by accidental mixture; it is not by the mere nursing of changes advantageous in the battle of life; it must be from the birth of some one couple, male and female, whose organisation is subjected to new conditions corresponding with each other, and having such force of self-continuance as to secure it against rever-

[1] Gould's "Trochilidæ," Introduction.

sion. It matters not how small the difference may be from the parent Form; if that difference be constant, and if it be associated with some difference equally constant in the female Form, it becomes at once a new Species. There are some cases mentioned by Mr. Gould which may possibly be examples of the first founding of a new Species. In the beautiful genus *Cynanthus*, he tells us that there are some local varieties near Bogota, in which the ornament is partially changing from blue to green; and it is a curious fact that this variation appears to be taking effect under the direction of some definite rule or "law,"—inasmuch as it is only the eight central feathers of the tail which are tipped with the new colour. Mr. Gould expressly says of one such variety from Ecuador, that it possesses characters so distinctive as to entitle it, in his opinion, to the rank of a separate Species. The very discussion of such a question shows the possibility of new births being the means of introducing new Species. But my object here is simply to point out that Mr. Darwin's theory offers no explanation of such births, either as respects their origin or their preservation, neither does it even approach to tracing these births to any physical law whatever. It fails also to recognise, even if it does not exclude, the relation which the birth of new Species has to the mental purpose of producing mere beauty and mere variety.

Nevertheless it may be true that ordinary generation has been the instrument employed ; but if so, it must be employed under extraordinary conditions, and directed to extraordinary results.

It will be seen, then, that the principle of Natural Selection has no bearing whatever on the Origin of Species, but only on the preservation and distribution of Species when they have arisen. I have already pointed out that Mr. Darwin does not always keep this distinction clearly in view, because he speaks of Natural Selection "producing" organs, or "adapting" them. It cannot be too often repeated that Natural Selection can produce nothing whatever, except the conservation or preservation of some variation otherwise originated. The true Origin of Species does not consist in the adjustments which help varieties to live and to prevail, but in those previous adjustments which cause those varieties to be born at all. Now what are these? Can they be traced or even guessed at? Mr. Darwin has a whole chapter on the Laws of Variation ;[1] and it is here, if anywhere, that we look for any suggestion as to the physical causes which account for the Origin as distinguished from the mere Preservation of Species. He candidly admits that his doctrine of

1 "Origin of Species," chap. v.

Natural Selection takes cognisance of variations only after they have arisen, and that it regards those variations as purely accidental in their origin, or, in other words, as due to chance. This, of course, he adds, is a supposition wholly incorrect, and only serves "to indicate plainly our ignorance of the cause of each particular variation." Accordingly, the Laws of Variation which he proceeds to indicate are merely, for the most part, certain observed facts in respect to Variation, and do not at all come under the category of Laws, in that higher sense in which the word Law indicates a discovered method under which Natural Forces are made to work. There is, however, in this chapter, one Law which approaches to a Law in the higher sense. Mr. Darwin, whilst candidly confessing our profound ignorance of the cause or origin of varieties, yet groups together a great class of facts as connected by a tie which he calls the "Correlation of Growth." Now what is this law—this observed Order of facts? It is, that variation in one part of an organism is, as a rule, accompanied with corresponding variations in other parts, and especially in those parts which are "homologous," that is to say, which occupy the same relative place in the general Plan.

This, however, is but a very imperfect definition of the vast Order of mysterious facts which are covered by the words, "Correlation of Growth." The fundamental idea which these words express is an Idea of wider and deeper significance in Nature, than Mr. Darwin seems to have perceived. There is a correlation between all natural organic growths; that is to say, that any variation of form in a single part has a constant relation to other variations of form in some other part or parts of the same organism. But "relation" is a vague word. There are many kinds of "relation"—there are indeed an infinite variety of kinds. What is the kind of relation that we detect in Correlated Growths? It is not until we ask ourselves this question that we discover what a deep question it is—how endless are the avenues of thought and of inquiry which it opens up.

First, one relation which we detect in all variations of organic growth, is simply the relation of symmetry. This kind and degree of Correlation of Growth prevails even in the world which we call Inorganic. The corresponding sides and angles of a crystal, for example, may be said to be correlated together. The nature of this relation is geometrical and numerical. It is a relation having reference to invariable rules of number. As regards its physical cause, all we can say is, that it is the result of forces whose property it is to aggregate the particles of matter in definite forms, which forms are symmetrical—

that is to say, they are forms having an axis with equal developments on either side. Correlation of Growth, therefore, in this sense points to the work of Forces, one of whose essential properties is Polarity—that is, equal and similar action in opposite directions. Now, this kind of Correlation of Growth may be traced upwards from simple Minerals through all the infinite complications of the organic world. It is unquestionably the basis of many of the Correlations of Growth prevailing in Plants and Animals. It is seen in the symmetrical arrangement of all vegetable and of all animal Forms. A central axis is traceable in them all; and the Bilateral or Radiated arrangement of their subordinate parts is one of the most fundamental and universal of all the Correlations of Growth.

This is one, but it is one only, of the Correlations of Growth which are constantly observed. It would lead us to expect that any change of form on one side of an animal would be accompanied by an exactly corresponding change on the other side: so that limbs on one side of the central axis, if changed at all, would change in exact and symmetrical accordance with the limbs on the opposite and corresponding side. This, accordingly, is one of the Correlations of Growth most constantly observed.

Now, it will be seen that Correlation of Growth, in

this first and simplest sense, runs alongside, as it were, of Correlation in another and higher sense. The relation between two equal and opposite growths, which is a relation, in the first place, of simple symmetry as between themselves, is always accompanied by another relation, in the second place, of correspondence or fitness as between these growths and external conditions. An organism which is developed unsymmetrically, unequally, would be not only ugly in its form, but it would be maimed and imperfect in its functions. Here, then, we see one kind and one idea of Correlation rising above another. Two growths might be correlated as regards each other, and might yet be wanting in any corresponding correlation of fitness and of function towards outward things. But the first of these two kinds of correlation would be useless without the last. And this last is obviously the higher and more complex Correlation of the two. It is higher, not only in the sense of being more complex, but as involving an idea which lifts us at once from a lower to a higher region of thought. Growths correlated as between each other according to mere symmetry of arrangement suggest nothing, except the work of Forces with inherent Polarity of action. But growths correlated with things outside the organism in which those growths occur,—and which can exert no physical effect upon it,—suggest at once the operation of Forces working under Adjustment with a view to Purpose.

When we see a Mineral salt crystallising under the power of a Voltaic Current, we see Correlation of Growth in its simplest form, and in visible connexion with its immediate cause. The particles of salt are marshalled in a constant Order—an Order, the principle of which is some central axis, with branches and branchlets grouped around it in equal and exquisite arrangement. Wonderful as this arrangement is, it suggests no other question to the mind than that which may be asked in respect to the ultimate nature and source of Polarity in Magnetic Force. But when we see two growths in an organism which not only are correlated to each other with reference to a centre, but are correlated also to external things with reference to Function, we see something which raises questions altogether different in kind. We have passed at once from the region of the What, and the How, into the region of the Why. The one kind of Correlation has reference to Physical Causes, the other kind of correlation has reference to those Mental Purposes which Physical Causes are made to serve. These two kinds of Correlation are perfectly distinct. They are as distinct as the correlation of equal pressures which a given volume of steam exerts upon the opposite sides of a boiler is distinct from the correlation between that pressure and its conversion into the driving-force of cranks and wheels, with all their adaptations for running on the rails, or for paddling in the sea. They are as distinct as the correlations of force developed in a Voltaic Battery are distinct from the adjustments which convert those forces into the means of communicating Thought.

Mr. Darwin has not pointed out this distinction clearly. Indeed, he does not seem to have had it in his view. He groups under one name,—the Correlation of Growth,—two classes of Phenomena, which are indeed always combined in fact, but which are entirely separate in idea. Correlation of Growth, in one sense, is that law of vital force which secures that any change in the shape of one limb in an animal shall be accompanied by a corresponding change in all the other limbs. Correlation of Growth in the other sense, is that adjustment of vital forces to the contingencies of external circumstance, which secures that all the changes which do take place shall be changes adapted to the discharge of new functions—to the fulfilment of new conditions of life—to command over new sources of enjoyment.

Keeping, then, clearly in our view the distinction between these two different kinds of Correlation of Growth, let us look at the phenomena actually presented in the aspect and history of Organic Forms, as respects both these kinds of Correlation.

As regards the first kind of Correlation, I have referred to the law of Bilateral Symmetry as the simplest and most obvious illustration. It is a law which at once connects itself with the idea of Polarity of Force. But though this be one kind of Correlation, almost universal, and may very probably be the foundation of every other, there are many Correlations of Growth between which and mere Polarity there is no visible connexion. The truth is that all the parts of an organism are bound together as one whole by a pervading system of correlations as intricate as they are obscure. When the organism is in health, and all its parts are working in harmony, the wonder of these correlations is not perceived. But they are brought out in a marked degree by the phenomena of disease, and also by the phenomena of monstrosity or malformation. The "sympathy" which the most distant and apparently unconnected parts of an organism show with each other, when one of them is affected by disease, is the index of correlations whose nature is utterly beyond the reach of our anatomy. It is the same with malformations. Mr. Darwin mentions one case of curious unintelligible correlation—viz., that a blue iris is associated in Cats with deafness; and, again, that the tortoise-shell colour of the fur is associated with the female sex in the same animal. In like manner the bright colours, and the more conspicuous ornaments of plumage in Birds, are correlated with the male sex. So likewise are vocal organs with the wonderful gift of song. In many insects the differences of form which are correlated with the differences of sex, are far greater than the differences which separate species and even genera. There are insects of which the male is a fly, whilst the female is a worm. There are many other cases of correlation between different growths in respect to which the nature and source of the connexion is equally unknown. For example, the complex stomachs of the Ruminant Order are uniformly associated with a particular form of hoof. Sometimes correlations the most constant and invariable are at the same time the most subtle and the most secret, because they are hid under other growths which are not so correlated, and which produce total diversities of outward aspect. One very curious class of correlations is the correlation between the internal structure of the teeth in animals, and the structure of other very distant portions of their frame. There lately was, for example, in the Zoological Gardens, a little animal, the Hyrax, not unlike a Rabbit in general appearance, and very like it in habit. It is the "Cony" of Scripture. Now this little animal will be found on examination to have limbs which do not terminate in a foot like a Rabbit, but in a divided hoof of peculiar form.

This hoof is in miniature like the hoof a Rhinoceros. If next we examine the teeth of the Hyrax we shall find that the materials of these teeth are also combined in the same manner, and after the same pattern as the teeth of the Rhinoceros. So it is with other parts of the same two animals. Along with the teeth and the hoofs there are certain other shapes of bones which seem to be under the same bond of likeness. Now these are Correlations of Growth between different parts of the same animal, and between the corresponding parts in two different species.

The conception, then, which we are led to form by this kind of Correlation between organic growths, is more complex than we had at first supposed. Mere Polarity of Force, leading to equal and opposite arrangement of subordinate parts, is not enough to satisfy the facts. This, indeed, may continue to be the type to which our thoughts refer, and by which we are helped to some more adequate idea of the facts. But the general impression left on the mind is this—that some One Force directs the form and structure of every organism, so that any change in one part of it is but the index of changes which run visibly or invisibly throughout the whole. The growths between which we detect a correlation, are not really separate things connected only by the few correspondences which we may be able to detect, but are part and parcel of one operation, the result of one Force, exerting its energies through channels which we cannot see, and according to laws of which we can form but a distant and faint conception. The truth is that Correlation in this sense is involved in the very word "Growth." Each part of every structure which is the result of growth must be correlated to every other part. This is essential to the very idea of growth, and to the very idea of an organism due to growth. When, therefore, Mr. Darwin says that one of the laws on which variation of form depends is Correlation of Growth, he simply says that variations of Growth depend on growth —for all growths must be correlated.

But Correlation in this sense helps but a little way indeed in conceiving the origin of a new Species. There might be the most minute and perfect harmony between the changes effected in an animal newly born without those changes tending even in the most remote degree towards the establishment of a new Form of Life. In order to that establishment there must be another correlation, and a correlation of a higher kind. There must be a correlation between those changes and all the outward conditions amidst which the new Form is to be placed and live. If this correlation fails the new Form will die. Yet, so far as we can see, this kind of correlation is without any physical cause. It is not necessarily involved, as the other kind of corre lation is, in the very idea of Growth. On the contrary, it is not only entirely separable in thought, but, as we see in monstrosities, it is sometimes separated in fact. We have no conception of any Force emanating from external things which shall mould the structure of an organism in harmony with themselves. Mr. Darwin freely confesses this, and says that many considerations "incline him to lay very little weight on the direct action of the conditions of life" in producing variety of Form. We can conceive, dimly indeed, but still we can conceive, how in the Humming Birds a special form of Wing shall be correlated with a special form of Bill. But we have no conception whatever how a special form of Bill should be correlated with a special form of Flower from which the Bill is to extract its food. Mr. Darwin has shown how an improved Bill, when once produced, will be preserved by finding external conditions to which it is adapted. But he has not shown, and he frankly confesses he has no idea, how the adapted variation of Bill comes to be born at all.

Yet it is this higher and more complex Correlation which is the most constant and the most obvious of all the facts of Nature. In these facts we see that the forces of Organic Growth are worked under rules of close adjustment to external conditions; and that particular shapes which might seem inseparably associated, if we looked at one Genus or one Family alone, are at once disjoined where different adaptations to Function are required. Let us take another example from the great Class of Birds. If we were to look only to the family of the *Anatidæ* (Ducks and Geese), we might suppose that there is a constant Correlation of Growth between webbed feet and spoon-shaped bills. But the real and efficient Correlation of Growth in this case is not between the spoon-bill and the web-foot, but between both of these and certain external conditions of life. The web-foot is correlated to an aquatic habitat: and the spoon-bill is correlated to spoon-food. And accordingly this association of form in foot and bill is at once dissolved where different external functions require a separation. In the Gulls, the Fulmars, and the Petrels, the web-foot is retained, because action upon the element of water is still required; but the correlated form of bill vanishes, and shapes altogether different are given,—shapes adapted, that is correlated, to different kinds of food, and to different methods of capture.

Again, there is another great family of Birds where some of the same forms are correlated with other forms entirely different, because of the different external Correlations which are required by Function. In the Divers the web-foot is mounted upon a flattened leg-bone, with the sharp edge set "fore and aft." Now what is this Correlation of Growth? It is, first, the Internal Correlation of those parts to each other, but secondly and principally, it is the External Correlation of both to their function of propelling under water. The form of the foot is correlated to the function of opposing the largest possible area of resistance to that medium, exactly where, for the purpose of swimming, the maximum of resistance is required; the knife-shaped leg-bone is correlated to the function of opposing the least possible resistance, precisely where, for the same purpose, the minimum of resistance is required. In Australia we have, in the *Ornithorynchus paradoxus*, the webbed feet correlated with the Duck-shaped Bill in an animal which does not belong to the Class of Birds at all.

There is another case of what may be called Correlated Correlations, which brings out very clearly the distinction which is so important in the philosophy of this great subject. Feathers are a kind of covering peculiar to the Class of Birds. Under every variety of modification they have one fundamental plan—a central shaft or quill to which lateral filaments are attached. Now there is a vast range of correlations between the different kinds of feather and the different Families or Species, and between different parts of the body in the same Species. But there are two Correlations of Growth in respect to feathers which are constant. In all cases, (excepting, of course, the Wingless Birds,) the feathers which grow from the fore-arm and finger-bones, constituting the Wings, are comparatively long, strong, tapering, elastic, and with thin lateral filaments, which filaments are closely hooked together by means of minute teeth fitting into each other, so that the whole shall form one continuous surface or web. This is a Correlation of Growth between one particular kind of feather, and one particular member of the body, which, in all Birds capable of flight, is constant, and amounts to a universal Law. Now let us contrast this with another Correlation of Growth which is equally constant. On the side of the head of all Birds, there is a patch of feathers of peculiar structure, with fine and slender shafts, and with the lateral filaments not hooked together as in the other case, but, on the contrary, always separated from each other—the whole series forming a fine and open network spread over the surface which they cover and protect. These feathers cover the orifice of the ear, and are called the auriculars. They are correlated with the curious passages, the finely hung clapper-bones, and all the elaborate mechanism of that organ. Such are the Internal Correlations. But they are intelligible only when considered in the light shed by other correlations which are external. The wing feathers with close continuous webs are correlated to the laws by which the passage of air may be prevented—the auricular feathers, with open unconnected webs, are correlated to the laws by which the passage of sound may be rendered easy. The one set of feathers are adapted to the active function of evoking and resisting atmospheric pressure by striking strong, yet light and elastic blows, upon the air—the other set of feathers are adapted to the passive function of allowing the free access of the waves of sound into the passages of the ear. These are but a few examples out of millions. Throughout the whole range of Nature the system of Internal Correlation is entirely subordinate to the system of External Correlation. Forms or growths which are inseparably joined with each other in one group of animals, are wholly divorced from each other in another group; whereas Forms which have correlations adapted to external conditions, are repeated over and over again across the widest gaps in the scale of Natural affinity.

If, then, it be true that New Species are created out of small variations in the form of Old Species, and this by way of Natural Generation, there must be some bond of connexion which determines those variations in a definite direction, and keeps up the External Correlations *pari passu* with the Internal Correlations. Natural Selection can have no part in this. Natural Selection seizes on these External Correlations when they have come to be. But Natural Selection cannot enter the secret chambers of the womb, and there shape the new Form in harmony with modified conditions of external life. How, then, are these external correlations provided for beforehand? There can be but one reply. It is by Utility, not acting as a Physical Cause upon organs already in existence, but acting through Motive as a Mental Purpose in contriving organs before they have begun to be. And where obvious utility does result, the only connecting Bond which can be conceived as capable of maintaining the Internal Correlations in harmony with the External Correlations, is the Bond of Creative Will giving to Organic Forces a foreseen direction. It is, in short, precisely the same bond which in all mechanism produces Structure in harmony with intended Function.

Hence it is that scientific men, in seeking expression for the ultimate ideas arrived at in the course of Physical research, find themselves compelled to borrow the language of Mechanical Invention. There is no other language which conveys an impression of the facts, or of the tie by which the facts are connected with each other. In the first chapter of this work I have had occasion to point out how true this is of Mr. Darwin's description of the Orchids, and of the curious functions of their structure. The correlations there are all external. But the same result appears in every other department of Science. In a remarkable paper on the "Constitution of the Universe," Professor Tyndall has occasion to speak of the non-luminous rays of heat emitted by all incandescent bodies,—rays which, though intensely hot, are altogether insensible to the eye. Now the Retina of the eye is a piece of mechanism whose Correlations are essentially External. It is the expansion of a special nerve whose function it is to be sensitive to certain particular vibrations, and to no other vibrations whatever. Light itself, therefore, is discovered to be merely a relative term—a word, in short, denoting nothing but an external Correlation between the Retina and vibrations of a certain kind and quality. Now what is the language which Professor Tyndall is constrained to use in explanation of facts so difficult of conception? It is the language of Mechanism, of mental Purpose and Design. "It is not," he says, "the size of a wave which determines its power of producing light; it is, broadly speaking, *the fitness of the wave to the Retina.* The ethereal pulses must follow each other with a certain rapidity of succession before they can produce light, and if their rapidity exceed a certain limit, they also fail to produce light. *The Retina is attuned*, if I may use the term, *to a certain range of vibrations*, beyond which, in both directions, it ceases to be of use." These are indeed wonderful Correlations which reveal to us fittings and adjustments of which we had no previous conception: but they give us no glimmering even, of knowledge as to the physical causes which have "attuned" a material organ so as to catch certain ethereal pulsations in the external world, and to make these the means of conveying to Man's Intelligence the enjoyment and the power of sight.

It will be seen, then, that when Mr. Darwin speaks of the Law of Correlation of Growth as a Law which determines variation in organic growths, he is really presenting to us under one phase two separate ideas which are radically distinct. One is the idea of different growths in the same organism, corresponding with each other in respect to arrangement,—or in respect to texture, or in respect to form,—or to some other point of comparison. The other idea is that these growths (each and all) correspond with the conditions of external nature in such a way as to fit them for the discharge of Function with some new adaptation, and consequently with some new advantage. In one aspect the Law of Correlation of Growth is (or at least may probably be) a Law in the strictest sense of the word; that is to say, the result of a Force acting according to its own definite modes and measures of operation. But the Law of Correlation of Growth in the other aspect, is a law only in the sense (1) of an observed order of facts; and (2) of that Order depending on Adjustment with a view to Purpose.

Many naturalists have spoken of the facts of organic likeness as sufficiently accounted for by referring them to Adherence to Type. Mr. Darwin complains that this phrase, as an explanation of organic likeness, is no explanation at all, but amounts only to a re-statement of the facts in another form of language. This is true; but it is equally true of his own phrase of Correlation of Growth.[1] "Adherence to Type" is not in the nature of a Physical Cause, but in the nature of a Mental Purpose. It is no explanation, therefore, to those faculties of the mind which seek for Methods of operation. In like manner "Correlation of Growth," in the only sense in which it is possible to connect it with the Origin of Species, is not a Physical Cause, but a Mental Purpose. The physical means by which that purpose is secured

[1] Mr. Wallace traces the whole Darwinian theory to six "general laws of the simplest kind—laws which," he emphatically adds, "*are in most cases mere statements of admitted facts.*" Again he says, "This series of facts or laws *are mere statements of what is the condition of nature.*"—*Journal of Science*, No. XVI. p. 472.

remain as dark as ever; and such of them as are conceivable by us, are seen, like all other physical forces, working to order, subject to direction, and having that direction determined by foresight, forethought, and contrivance.

Correlation of Growth, in the sense of external adaptations, may be said to be the most universal of all the Laws of Nature. But it accounts for the Origin of Organic Forms only in the same sense in which it accounts for the origin of all other phenomena, which in their result exhibit adaptations, or fittings into use and service. Let us take, as an example, the origin, nature, and capacities of Coal. That substance is correlated in a truly wonderful manner with the needs, the powers, and the capacities of Man. It contains within itself, in a form condensed and portable, a store of physical Force of incredible amount. The particles of one pound weight of it are held together by a Force which, when liberated and applied in the form of heat, is capable of lifting one million times its own weight to the height of one foot.[1] No other substance known to Man is to be compared with this as a furnisher of Force. This is its function in the world. It is a function relating to Man's mechanical and inventive powers; and coal has been rendered

[1] "The Coal Question."—W. S. Jevons, 1865.

capable of discharging this function by processes of preparation which began millions of ages before Man was born. But these External Correlations are a result arising by way of natural consequence out of certain physical causes working to order, that is to say, out of Internal Correlations of Growth between Solar Heat and Vegetable Structure, and again between these and the causes which occasion interchange between sea and land. No explanation so definite as this can be given of the method in which Vital Forces are made to evolve a new Form of Life. But even if such explanation could be given, it would render no account at all of the fittings of that Form into the outward requirements of its life. These are Correlations which in their very nature belong to Mind, are the work of Mind, and are intelligible only in the light of Mind.

I do not represent this conclusion as one necessarily adverse to Mr. Darwin's Theory on the Origin of Species. It is a conclusion which he would probably be willing to accept. I only desire to point out in how very limited a sense that Theory can be said to trace Creation to a "Law" at all, and how entirely inadequate that Theory is to account by any physical cause for the Origin of Species.

The only senses, therefore, in which we get any glimpse of Creation by Law are these—*1st*, That the close physical connexion between different Specific Forms is probably due to the operation of some Force or Forces common to them all; *2d*, That these Forces have been employed and worked, with others equally unknown, for the attainment of such ends as the multiplication of Life, in Forms fitted for new spheres of enjoyment, and for the display of new kinds of beauty.

Is there anything in this conclusion to conflict with such knowledge as we have from other sources of the nature and working of Creative Power? I do not know on what authority it is that we so often speak as if Creation were not Creation unless it work from nothing as its material, and by nothing as its means. We know that out of the "dust of the ground"—that is, out of the ordinary elements of Nature—are our own bodies formed, and the bodies of all living things. Nor is there anything which should shock us in the idea that the creation of new Forms, any more than their propagation, has been brought about by the use and instrumentality of means. In a theological point of view it matters nothing what those means have been. I agree with M. Guizot, when he says that "Those only would be serious adversaries of the doctrine of Creation who could affirm that the universe—the earth, and Man upon it—have been from all eternity, and in all respects, just what they are now."[1] But this cannot be affirmed except in the teeth of facts which Science has clearly ascertained. There has been a continual coming-to-be of new Forms of Life.[2] This is Creation, no matter what have been the laws or forces employed by Creative Power.

The truth is, that the theory which fixes upon Inheritance as the cause of organic likeness, startles us only when it is applied to Forms in which unlikeness is more prominent than resemblance. The idea, for example, that the different kinds of Pigeon, or of Humming Birds, have all descended through successive variation from some one ancestral pair, whether it be true or not, would not startle any one. Yet, if this be true, we must be prepared for the same surmise extending farther. The advocates of Development urge that Time is a powerful factor. They say that if changes small, but constant enough, and definite enough, to constitute new Species, can and do arise out of born varieties, it is impossible to fix the limits of divergence which may be reached in the course of ages. It does not follow, on the other hand, that there is no such limit

[1] "Méditations sur l'Essence de la Religion Chrétienne," p. 49.

[2] "We discern no evidence of a pause or intromission in the creation or coming-to-be of new plants and animals."—*Instances of the Power of God as manifested in His Animal Creation, by Professor Owen.*

because we cannot fix it. It does not necessarily follow that because we admit the idea of the Rock-dove, and the Turtle-dove, and the Ring-dove being all descended from one ancestral Pigeon, we are bound to accept the idea of the Whale, and the Antelope, and the Monkey being all descended from some one primeval Mammal. Mr. Darwin says, truly enough, that Inheritance "is that cause which alone, as far as we positively know, produces organisms quite like, or nearly like, each other." But this is no reason why we should conclude that Inheritance is the only cause which can produce Organisms quite unlike, or only very partially like each other. We are surely not entitled to assume that all degrees and kinds of likeness can arise only from this single cause. Yet until this extreme proposition be proved, or rendered probable, we have a sound scientific basis for doubting the application of the theory, precisely in proportion to the unlikeness of the animals to which it is applied.

And this is the ground of reasoning, besides the ground of feeling, on which we revolt from the doctrine as applied to Man. We do so because we are conscious of an amount and of a kind of difference between ourselves and the lower animals, which is, in sober truth, immeasurable, in spite of the close affinities of bodily structure. Yet the closeness of these affinities is a fact; and it may with truth be said that in contrast with the gulf of separation in all resulting characters, these affinities are among the profoundest mysteries of Nature. Professor Huxley, in his work on "Man's Place in Nature," has endeavoured to prove that, so far as mere physical structure is concerned, "the differences which separate him from the Gorilla and the Chimpanzee are not so great as those which separate the Gorilla from the lower Apes." On the frontispiece of this work he exhibits in series the skeletons of the Anthropoid Apes and of Man. It is a grim and grotesque procession. The Form which leads it, however like the others in general structural plan, is wonderfully different in those lines and shapes of Matter which have such mysterious power of expressing the characters of Mind. And significant as those differences are in the skeleton, they are as nothing to the differences which emerge in the living creatures. Huxley himself admits that these differences amount to "an enormous gulf," to a "divergence immeasurable—practically infinite." What more striking proof could we have than this, that Organic Forms are but as clay in the hands of the Potter, and that the "Law" of Structure is entirely subordinate to the "Law" of Purpose and Intention under which the various parts of that structure are combined for use?

But Science will continue to ask, even if she never gets an answer, What is the community of physical cause which produces this community of resulting structure? The fact which it is most difficult to disengage from the theory of Development, or, in other words, from the theory of Creation by Birth, is the existence of rudimentary or aborted organs; the existence of teeth, for example, in the jaws of the Whale—teeth which never cut the gum, and which are entirely useless to the animal. We have an inherent conviction that this must have some use in the future,—that is, in some organism to be born from this one,—or else it must have had some use in the past,—that is, in some organism from which this one has descended. In either case the power of Inheritance is suggested to the mind. We think instinctively of the existence of some Derivative Form in which these teeth have been, or are to be turned to use. It is only fair towards the Theory of Creation by Birth, to admit that it does explain the existence of useless organs in a sense in which no other Theory explains them. It would be almost a necessary consequence of Creation by Birth, that there must be stages in which the ultimate use of new Forms could not be yet apparent. And if mere beauty or variety were in themselves objects which Creative Power sets before itself, then, also, we might expect to meet with modifications of structure having no other apparent use. Both these explanations, however, exclude Mr. Darwin's idea of Natural Selection; because this is a process which can never operate, except through the agency of actual use and disuse, upon organs already existing and capable of discharging function. The only theory of Creation by Birth which really does afford some explanation of the facts, is a theory which assumes modifications of structure to be entirely independent of the effect of actual use or disuse. Mr. Darwin himself candidly admits that in flowers, at least, the forces of Correlated Growth do "modify important structures independent of Utility, and therefore of Natural Selection." This admission must be extended to all organic growths. There must have been a time with all of them when they began to be; and, therefore, a time before Natural Selection had room to play. These considerations, however, only serve to put a higher interpretation on the Theory of Creation by Birth. They do not condemn it.

One suggestion, indeed, has been made on this subject which I think it is impossible to accept. When men were yet unwilling to admit the existence of life and death upon the globe so long before the creation of Man, it used to be said that fossils were only "sports of nature." So in our own day, I have heard it said that rudimentary organs are merely intended to satisfy that condition of our finite minds in virtue of which we are unable to conceive Creation, except in connexion with some History and Method of growth. And so, as a condescension to this weakness, aborted members are given to suggest a History which was never true, and a Method which was never followed! Now of one thing we may be sure, that there are no fictions of this kind in Nature, and no bad jokes. Whatever natural things really point to, they point to faithfully; and the conclusions really indicated are never false. Abortive organs mean something, and they mean it truly.

Still, there is no proof that Inheritance is the only cause from which such structures can arise. In the inorganic world we know that not mere similarity, but absolute identity of form, as in crystals, is the result of laws which have nothing to do with Inheritance, but of forces whose nature it is to aggregate the particles of matter in identic shapes. It is impossible to say how far a similar unity of effect may have been impressed on the forces through which vital Organisms are first started on their way. There are some essential resemblances between all Forms of Life which it is impossible even in imagination to connect with community of blood by descent. For example, the Bilateral arrangement is common to all Organisms, down at least to the Radiata, and in this great class we have the same principle of Polarity developed in a circle. Again, the general mechanism of the digestive organs by which food is in part assimilated and part rejected, is also common through a range of equal extent. Indeed, it may be said with truth, that never in all the changes of Time has there been any alteration throughout the whole scale of Organic Life, in the fundamental principles of chemical and mechanical adjustment, on which the great animal functions of Respiration, Circulation, and Reproduction, have been provided for.[1] These are fundamental similarities of plan, depending probably on the very nature of Forces which necessitate these adjustments in order to the production of the phenomena of Life—Forces of which we know nothing, but which we have not the slightest reason to suppose to be due to Inheritance. Other similarities of plan may depend on the same laws, equally unconnected with Inheritance by descent.

Inheritance, indeed, has been suggested as the cause of organic likeness, mainly because there is a difficulty in conceiving any other. But there is at least an equal difficulty in conceiving the applicability of this cause to Man. We have already seen[2] that M. Guizot lays it

[1] Agassiz' "Geological Sketches," p. 41. London, 1860.

[2] *Ante*, page 28.

down as a physical impossibility that Man—the human pair—can have been introduced into the world except in complete stature—in the full possession of all his faculties and powers. He holds it as certain that on no other condition could Man, on his first appearance, have been able to survive and to found the human family. Even those who question whether this argument is entitled to the rank of a self-evident physical truth, must admit that it is at least quite as good as the opposite assertion, that any origin except the origin of natural birth is inconceivable. Where our gnorance is so profound, no reasoning of this kind is of much value. There is undoubtedly much to be said in support of M. Guizot's position. Certainly, Man as a mere animal is the most helpless of all animals. His whole frame has relation to his mind, and apart from that relation, it is feebler than the frame of any of the brutes. All its members are Correlated amongst each other with the functions of his Brain, so that action may follow upon Knowledge—so that embodiment may be possible to Thought. Yet in its plan and structure his frame is homologically, that is ideally, the same as the frame of the brutes—organ answering to organ, and bone to bone.

The words "Adherence to Type" are words expressive of an Idea, of a Purpose, which we see fulfilled in

Organic Forms. But this purpose must have sought its own accomplishment by the use of means, and the question of Science always is, what were these means? Love of beauty is equally a Purpose which we see fulfilled in Nature, but in the case of the Humming Birds this has been accomplished by giving to their plumes the structure of "Thin Plates,"—a structure which decomposes light and flings back its prismatic colours to the eye. Fitness and special adaptation is another of the purposes of Creation, but this also is attained through the careful arrangement, and pliability to use, of physical laws. In like manner, "Adherence to Type" is the expression of a fact, or the statement of a Purpose, which, like all the other purposes fulfilled in Nature, invites to an investigation of the instrumentality employed. We see the Purpose, but we do not see the Method. We see the purpose, for example, in the wonderful adaptability of the Vertebrate Type to the infinite varieties of Life to which it serves as an organ and a home. Science should be allowed without suspicion or remonstrance to pursue her proper object, which is to detect, if she can, what the method of this work has been. There is no point, short of the last and highest, at which Science can be satisfied. Her curiosity is insatiable. It is a curiosity representing man's desire of knowledge. But that desire extends into regions where the means of investigation cease, and in which the processes of Verification are of no avail. Above and behind every Detected Method in Nature there lies the same ultimate question as before—What is it by which this is done?

It is the great mystery of our Being that we have powers impelling us to ask such questions on the history of Creation, when we have no powers enabling us to solve them. Ideas and faint suggestions of reply are ever passing across the outer limits of the Mind, as meteors pass across the margin of the atmosphere, but we endeavour in vain to grasp or understand them. The faculties both of reason and of imagination fall back with a sense of impotence upon some favourite phrase—some form of words built up out of the materials of analogy, and out of the experience of a Mind, which, being finite, is not creative. We beat against the bars in vain. The only real rest is in the confession of ignorance, and the confession, too, that all ultimate physical Truth is beyond the reach of Science.[1] It is

[1] I have slightly altered this passage as it stood in the earlier editions, because, although the context clearly indicates its reference to Physical truth, it has been quoted by Mr. Lewes as granting all that the Positive Philosophy demands. There is a sense, of course, in which it may be said that no Truth knowable by man can be "ultimate." That is to say, there is no Truth even conceivable, respecting which we might not ask, or desire to ask, farther questions. But there is no use in appearing to agree with those

probable that even the nearest methods of Creation, though far short of ultimate truths, lie behind a veil too thick for us to penetrate. It is here surely, if it is anywhere in the sphere of natural investigation, that the Man of Science may lay down the weapons of his analysis, and say, "I do not exercise myself in great matters, or in things which are too high for me."[1]

There is at least one conclusion which is certain, namely, this—that no theory in respect to the means and method employed in the work of Creation—provided such theory takes in all the facts—can have the slightest effect in removing that work from the relation in which it stands to the attributes of Will. All such theories are, and can only be "simply questions of how the Creator has worked." This is the confession made in respect to Mr. Darwin's theory by one of the most competent of its supporters.[2] Creation by Law—Evolution by Law—Development by Law, or, as including all those kindred ideas, the Reign of Law, is nothing but the reign of Creative Force directed by Creative Knowledge, worked under the control of Creative Power, and in fulfilment of Creative Purpose.

from whom in reality I so widely differ. The definition of Truth which Mr. Lewes would consider "ultimate," and therefore unattainable, is very different from the definition which I should give, and have given.

[1] Psalm cxxxi.

[2] Mr. Wallace.—*Journal of Science*, No. XVI. p. 473

CHAPTER VI.

THE REIGN OF LAW IN THE REALM OF MIND.

WHEN we pass from the phenomena of Matter to the phenomena of Mind, we do not pass from under the Reign of Law. Here, too, facts do range themselves in an observed Order: here, too, there is a chain of cause and effect running throughout all events: here, too, we see around us, and feel within us, the work of Forces which have always a certain definite tendency to produce certain definite results: here, too, it is by combination and adjustment among these Forces that they are mutually held in check: here, too, accordingly, special ends can only be accomplished by the use of special means.

But then the question immediately occurs to us—can we speak of Law, or of Force, or of "cause and effect," as applied to the phenomena of Mind, in the same sense in which we speak of them as applied to the phenomena of Matter? Is it only by distant analogy, or as expressing ideas really the same, that we use the same terms of both?

Undoubtedly the first thought which suggests itself to the mind is, that a material Force and a moral or intellectual Force are essentially different in kind,—not subject to conditions the same, or even similar. But are we sure of this? Are we sure that the Forces which we call Material are not, after all, but manifestations of mental energy and Will? We have already seen that such evidence as we have is all tending the other way. The conclusions forced upon us have been these:—first, that the more we know of Nature the more certain it appears that a multiplicity of separate forces does not exist, but that all her forces pass into each other, and are but modifications of some One Force which is the source and centre of the rest: secondly, that all of them are governed in their mutual relations by principles of arrangement which are purely mental: thirdly, that of the ultimate seat of Force in any form we know nothing directly: and fourthly, that the nearest conception we can ever have of Force is derived from our own consciousness of vital power.

If these conclusions be true, it follows that, whether as regards that in which Force in itself consists, or as regards the conditions under which Force is used, it need not surprise us if in passing from the material world to the world of Mind, we see that Law, in the same sense, prevails in the phenomena of both. But as this is a subject of much difficulty, and of much importance, it may be well to examine it a little nearer.

The first and most palpable form in which we see that Mind is subject to Law, is in its connexion with the Body. And this connexion is so close that we know neither where it begins nor where it ends. The extent and nature of it can be known only by the same kind of reasoning and observation by which we attain to any knowledge of the external world. For indeed our Bodies seem part of the external world to us. We see their form as we see the form of other things, but we do not see their structure, neither do we feel it, nor can we arrive at it, except as a matter of obscure and difficult research. It is literally true that some of the most distant objects in the Universe are more accessible to our observation, and in some respects more intelligible to our understanding, than the material frame in which we live. Man had discovered much concerning the circulation of the Planets before he had discovered anything concerning the circulation of his own blood.[1] Yet so near is the current of that blood to him, so much is it a part of himself, that when it stops, in an instant "all his thoughts perish." Nevertheless, the Mind is not conscious of its own dependence on material organs. Even

[1] Kepler and Harvey were contemporaries; but Copernicus had preceded them by nearly a hundred years.

in respect to those exertions of the Will which are expressed in movements of the Body, we are conscious only of the Will, and of the Will being exerted with success; but we are entirely unconscious of the machinery which intervenes between the intention and the accomplishment of the act intended. Such movements of the Body appear to us as if they were direct acts of Will. Yet nothing can be more certain than that the communication is not direct but indirect—and even elaborately circuitous. It is only when the ropes and pulleys are broken that we discover how that which we call our Will can only run in appointed channels—which channels are material, and are laid down upon a plan, like conducting wires, as if for the conveyance of a material Force.

Nor does it end here—this close connexion between Mind and Matter. So far from being less close, it seems to be only closer and closer when we pass to mental operations in which no apparent movements of the Body are concerned. In the exercise of pure Reason, in passing from one mental conception to another, when by an effort of our Will we turn our attention to a new question, and in the twinkling of an eye pursue a fresh train of thought,—above all, when our affections go forth towards those who are the objects of them—in all these operations, if anywhere, we feel as if we were free from mechanism—from "organs"—from Matter in any form.

So it seems till we are brought face to face with the terrible phenomena of disease. Then our delusion is dispelled, and we know how frail we are. Then we find that the same stroke which paralyses the movement of a limb, may paralyse, not less effectually, all the powers of Reason, of Memory, and of Will. And the Affections,—what becomes of them? These too, which seem so purely spiritual, we find out to be dependent on material structure. Every physician knows that a frequent consequence of cerebral disease is a total change of character. There is no symptom of insanity more common than the growth of dislike and aversion to those who, in health, had been the most loved on earth. Change of every kind and degree in the character and structure of Mind is the immediate result of corresponding changes in the structure and substance of the Brain. The pure may become impure; the loving may become malignant; the simple-minded may become suspicious; the generous may become engrossed with self; the strong-minded may become imbecile,—the whole man may be broken down, and may live for years without consciousness and without emotion. How painfully does the Brain sometimes indicate its functions! What is it in the aspect of Idiotcy, in many of its forms, which we instantly recognise, and never can mistake? In that low, pinched, and retiring brow, we see instinctively that

Reason cannot hold her seat. These facts do not stand alone. Not only are there some parallel facts, but all the living world is full of them. The whole range of animal creation, from Man down to the Reptile and the Fish, testifies to the universal law of an ascending scale of mental capacity being coincident with an ascending degree of cerebral organisation. No series of facts, tending to the establishment of any physical truth, is more complete or more conclusive than the chain which connects the functions of the Brain with the phenomena of Mind.

But here, again, let us beware of the fallacies which may arise from a failure to recognise the exact import of the words we use. In the ears of many it sounds like Materialism to say that Thought is a function of the Brain. But it has been already shown in a previous chapter that Function is merely the word by which we describe that work which any given piece of mechanism has been adjusted to perform. The Power, or Force, which is developed by means of an "organ," is not identical with that organ, nor with any of its parts, nor with the materials of which it is composed, nor even with its mechanism as a whole. It does not follow, for example, that Electricity is identical with the tissues of a fish, because it is developed out of the battery of a Torpedo or a Gymnotus. Yet it is true that the development and discharge of Electricity is the "function" of those Fish-organs:—that is to say, this is the work which they have been adjusted to perform. Still less do we confound Thought with Brain when we acknowledge the fact that Brain in our Organism is inseparably connected with the power of thinking.

Yet inferences as false as this, and very nearly related to it, have actually been drawn by eminent men from the facts of cerebral action. Thus it has been declared that a knowledge of Brain, under a name which is in itself a fallacy — Phrenology — is the only sure foundation of Mental Science. This is a mere confusion of thought, even if the phrenological mapping of the Brain were as certainly correct as it is really doubtful. That particular faculties of the Mind may be connected with particular portions of the Brain, is not in itself more difficult to understand or to believe than that the Mind, as a whole, is connected with the Brain as a whole. Whether it be so or not is a question purely of observation and of fact. But this, at least, is certain,—that the different faculties and affections of the Mind must be discriminated from each other before it is possible to assign to them a local habitation. The Mind must be mapped first, and then its Organ. No additional knowledge is given to us of any one mental faculty, by proving that it is connected with some special bit of the mysterious substance of which that organ is composed. Love is Love, and nothing else; Hatred is Hatred, and nothing else; Reverence is Reverence, and nothing else; the pure intellectual perception of a Logical Necessity is itself, and nothing else;—however clearly it may be proved that each of these is a function of some separate region of the Brain. When the Phrenologist, taking in his hand a human skull, and lifting its upper cover, tells us that the oval of convoluted matter which is thus exposed to view "manifests the moral sentiments," what light does he throw on these? The moral sentiments!—what do these include? The power of seeing Moral Beauty, and of loving Truth—the sense of Justice, and the desire of serving in her cause—Conscience and Benevolence, Charity and Faith—all that is best and noblest in the human spirit—these are "manifested" in that bit of Matter! What new information does this give us on the nature or the office of those glorious attributes which are the joy of Earth and Heaven? None at all. They are just what we knew them before to be.

Phrenology is no longer popular, as it once was, among Physiologists. Its mapping of the Brain is now generally admitted to be imaginary. But the fundamental error of the Phrenological School did not lie merely, or even mainly, in any mistake as to the mapping of the Brain. It lay in the idea that a Science of Mind can be founded in any shape or form upon the discoveries of anatomy. Their error lay in the notion that Physiology can ever be the basis of Psychology. And this is an error, and a confusion of thought, which survives Phrenology. A profound interest indeed attaches to every new fact which connects together the parallel phenomena of Mind and of Organisation. But it is the phenomena of Mind, and it is these alone, of which we are directly cognisant, and it is from these that we must start as the basis of all Psychological research. This is true even of those phenomena of the mind which are most purely animal. Sensation, for example, may be traced with absolute demonstration to certain nerves. This may throw a new light on the method by which Sensation is rendered possible; but it throws no new light whatever upon what Sensation is. It is that which we know and feel it to be, and it is neither more nor less since the knife of the anatomist has laid bare the channels along which it comes. Still more is this true of the Intellectual Powers. Yet there are Philosophers who appear to think that some new light is cast upon Sensation when they call it an affection of the "Sensory Ganglia;" that Thought is in some measure explained when it is called "Cerebration," and that the Laws of Intellect are reduced to scientific expression when they are described as the working of the "Cerebral Ganglia."

all this is a mere idle play on words. It is an attempt to put that first which must be last, and that last which must be first. The general fact of the dependence of Mind on a Bodily organisation is a fact which contains within itself all the lesser facts of Physiological discovery. They are not, and they cannot be, new in kind. They do not even help us to conceive how, through any mechanism, the power of Thought can be evolved. Still less do they give us any new view of that which Thought, in itself, is.

This connexion, therefore, between Mind and Brain, although it is a universal "law" of our being, is a law recognised by us only in the sense in which Law is applied to "an observed Order of facts." But like every other Order of this kind, it implies a Force or an arrangement of Forces out of which the Order comes. It implies, too, that this arrangement of Forces is necessary to the evolution and play of mental faculties in the form in which they are possessed by us. Consequently these faculties are seen taking their place among all the other phenomena of the world. They are seen to be under the Reign of Law in this largest and highest sense of all—that they depend upon Adjustment, and that adjustment so delicate that the slightest disturbance of it deranges the whole resulting phenomena of Mind. Mind, as developed in us, has its very existence and working dependent on imperative physical conditions, which conditions are met only by elaborate contrivance.

We have no knowledge what the Forces are which demand this obedience, and which call for this contrivance. We have even an insuperable difficulty in conceiving what they can be. It almost seems as if there were a barrier in the very nature of our minds against the possibility of conceiving how any combination of material forces can either result in Mind, or can be necessary to the working of its powers, or can be concerned even in giving it an abode.[1] "We cannot conceive," says Dr. Andrew Combe, "even in the remotest manner, in what way the Brain—a compound of water, albumen, fat, and phosphate salts—operates in the generating of Thought." And yet there is one experience which brings the fact of this close connexion within the direct recognition of Consciousness. We know and feel that the act of severe thinking is attended with the expenditure of Force. The close, steady, continuous application of the mind to any subject requiring the exercise of our higher intellectual

[1] "Aperta simplexque mens, nullâ re adjunctâ quæ sentire possit, fugere intelligentiæ nostræ vim et notionem videtur."—Cicero, "De Nat. Deor." lib. x. c. 11.

This is true only in one sense. It is very far from being true, that the connexion between Mind and Matter is a necessity of thought.

faculties, is well known to be "hard work." Without causing any bodily movement of which we are conscious, it produces, nevertheless, bodily exhaustion. It occasions the expenditure of a physical force, or at least of a force for which we have no other name. It is not uncommon for men of great age to be able to exert undiminished powers of mind for one or two hours, and then to lapse into comparative imbecility. Thus the exertion of the Brain is like the exertion of a muscle, and is attended with the same effects. There is fatigue; and with excessive fatigue the power of motion stops.

Yet such facts as these only puzzle us—they do not help us to any clear idea of the nature or manner of a connexion which is indeed incomprehensible. We know of Mind only as itself, and as nothing else. The difference between it and all other things seems infinite and immeasurable. No doubt this difficulty, or at least part of it, arises not from any misconception as to what Mind is, (for of this our knowledge is direct,) but from a misconception as to what Matter is—and what the Forces are which we call material. Close analysis of the phenomena of Nature, and of our own ideas in regard to them, has already prepared us to believe, that these Forces which work in Matter and produce in us the impressions from which we derive our conceptions of it, are themselves immaterial, and can be traced running up into a region where they are lost in the light of Mind. The Christian doctrine of the Resurrection of the Body sanctions and involves the notion that there is some deep connexion between Spirit and Form which is essential, and which cannot be finally sundered even in the divorce of Death. The affections hold to this idea even more firmly than the intellect. Hence the noble and passionate exclamation of the Poet—

> "Eternal Form shall still divide
> The Eternal Soul from all beside,
> And I shall know him when we meet."[1]

But this first sense in which Mind is under the Reign of Law—that is, its dependence on the Body, prepares us for yet other senses in which it lies under the same dominion. The very fact that the Mind is itself unconscious of its dependence upon Matter, and of the manner and conditions of its connexion with "organs," teaches us that there is a large class of phenomena connected with Mind, of which we should be entirely ignorant if we trusted to the direct evidence of Consciousness alone. This ought not to inspire us with any distrust of Consciousness in those matters in which it is a competent and indeed the only witness. But there is a large class of phenomena of which Consciousness properly so

[1] Tennyson's "In Memoriam," No. xlvi.

called, that is, the direct perception of the Mind of its own present workings, does not inform us. The Mind looking in upon itself sees itself only, and does not see either the mechanism through which it is able to work at all, nor many of the forces which operate in it and upon it. These, some of them at least, can only be arrived at by the same processes of reasoning and observation which we apply to the external world, and by which we ascertain the action and reaction of involuntary agents.

There is nothing of which it is so difficult to persuade ourselves as of this. In the apprehension of Consciousness the sense of Will is so strong within us that it blinds us to the insuperable conditions which limit both what we will and what we do. That our Wills, of whose freedom we are conscious, should often be determined by influences of which we have no consciousness at all; that our opinions should as often be the result of causes and not of reasons; that our actions should follow a course marked out by conditions which we fail to recognise as having any determining effect upon them—these are conclusions against which we are apt to rebel—as depriving us of a part of our free and intelligent agency. Hence the indignation with which men resent being told that they have been impelled by motives other than the motives which are avowed, and other than the motives which are consciously entertained. Yet the fact of their being so impelled is often perfectly plain to those around them. The reply, however, is always ready: "You seem to know my motives, and the causes of my conduct better than I know them myself,"—as if the proposition so stated were evidently absurd. But it is, on the contrary, a proposition which may well be true. Bystanders very often see the forces telling upon our Will much more clearly than we see them ourselves. It is possible, indeed, by a vigorous effort of self-analysis to see all that others see, and a great deal more. Those who are able really to look in upon themselves, can often detect the influences which have been acting on their minds, colouring their opinions, and determining their conduct in a degree which the higher faculties would be glad to disown and disavow. There is nothing more wonderful in the constitution of our minds than the power we have of standing aside, as it were, for a time, from the ordinary channel of our own thoughts, and of looking back upon their currents coming down from the hills of Memory and Association to join their issues in our present life. But this sort of looking in upon ourselves, and treating ourselves as a subject of natural history, is to all men a difficult, and to most men an impossible, operation. They have neither time for it nor thought for it. The conscious energies of the Will are so near us, and so ever present with us, that they shut out our view of the forces which lie behind. Yet there are some facts common in the experience of all men which may help us to a conception of the truth. One of these is the fact of Mind growing with the growth of years—a fact determined by the recollection of childhood, of youth, and of maturity. By comparing ourselves with ourselves at former periods of life—by the memory of feelings, and of opinions, and of methods of thought which we have outgrown and left behind us, we can detect the action of forces which have told upon our minds—traces, in short, of the laws to which they have been subject. Some of these laws have been nothing more than laws of physical growth —the conceptions of the Mind undergoing a development consequent on the growth of our material Organism.

Another fact bearing on the same question, but which is more easily observed in others than in ourselves, is the frequent determination of mental qualities by hereditary transmission. The famous question, as to the Origin of our Ideas, and how far they are due respectively to Experience, to Association, or to Intuition, has been discussed by Metaphysicians with far too little reference to the organic phenomena which are so closely related to the phenomena of Mind. It is not true, indeed, that Psychology is subordinate to Physiology; but it is true that these two are so intimately connected, that neither is independent of the other. Man is not a disembodied Spirit, but a Being whose mental powers are subject to the laws of a material organisation. And so it is that almost every fact in Physiology has an intimate bearing on some question or other in the Philosophy of Mind. No better illustration could be given than one which arises out of this question of the Origin of our Ideas. In one of the many formulæ of expression to which Mr. J. S. Mill has reduced the assertion that Experience is the source and origin of all our thoughts and actions, he is obliged to except from the sweep of that assertion the voluntary movements of the Body. He says, "We bring about any fact, *other than our own muscular contractions*, by means of some other fact which *experience has shown* to be followed by it."[1] Now let us observe the immense significance which attaches to this exception. Why is Mr. Mill compelled to make it? Because he mixes up in one assertion two propositions which are totally distinct, one being true universally, and the other being true only partially. The first proposition is, that all facts which we can "bring about," must be so brought about by

[1] "Auguste Comte and Positivism," by J. S. Mill, p. 7.

the use of means. This is true universally. The second proposition is, that we are guided to the knowledge of those means by Experience alone. Now, this last proposition is not true, as Mr. Mill is obliged to confess, of the whole class of facts which are brought about by vital effort. But the muscular contractions of the Body are no exception whatever to the mere general affirmation that all actions must have a cause, or in other words, must be brought about by the use of means. Exceptions they are, however, to the affirmation that the nature of those means is made known to us by Experience. The sentence, in so far as it asserts the universal Law of Causation, might have been so framed as to require no abatement or exception whatever. "We bring about any fact by means of some other fact which we know either by experience *or by Intuition* to be followed by it." In this form the sentence is absolutely true, and applies to "our own muscular contractions," as well as to every other action. But philosophers who support the doctrine of Experience do not like the word "intuition;" and though they cannot do without it altogether, they use it as seldom as they can. They feel very naturally, and very truly, that if Intuition be admitted in regard to the ultimate phenomena of Volition, the idea will not easily be dispelled that Intuition may extend also to the ultimate phenomena of

Thought. Now the muscular contractions of the Body stand at the very fount and origin of all we do; and it is more than probable that analogous movements of the Brain stand as near to the origin of all we think.

The bearing of this question on the Philosophy of Mind cannot be mistaken. The muscular contractions of the Body are of two kinds—one kind is constant, automatic, and lasting with the duration of life itself. The other kind is intermittent, voluntary, and capable of being destroyed whilst the Consciousness, and the Intelligence, and the Will are still in use. Both these kinds of action are rendered possible by the use of means: but it is only in the case of one of them that those means are placed at the bidding of the Will. Yet it is not Experience which teaches us how to use those means. It is purely Instinct or Intuition. We are not even conscious of the very existence of the means which we employ, and the profoundest researches of Science do not even yet give us the faintest notion what their ultimate nature is. No experience whatever is required to teach a child how to extend its limbs or how to exert its voice. Nevertheless, neither of these things can be done except through the use of means. The only difference between these actions and actions of a more complicated kind is, that the appropriate means are resorted to and employed by Intuition.

The Will which moves the limbs, and moves them through the use of a complicated machinery, is born with the Organism of which that machinery forms a part, and has an instinctive knowledge how to use it. Now, it is against the analogy of Nature to suppose that this great class of facts respecting the powers of the Body are without some corresponding facts respecting the powers of Mind. Indeed, all vital phenomena of this kind are in themselves necessarily phenomena both of Body and of Mind. The close connexion which exists between the two, and the inseparable analogies which unite all their workings, render it therefore almost certain that the Mind is to be regarded as having both kinds of movement which the physical Organism possesses—that is, faculties which are automatic in their action—and other faculties which, though subject to direction by the Will, yet work upon the materials presented to them in a manner strictly intuitive and independent of all experience.

And as the abnormal phenomena of disease, or of malformation, often throw an important light on the structure of the body, so do certain abnormal intellectual phenomena give us strange glimpses occasionally into the powers of Mind. Among those phenomena, none are more curious than the intuitive powers of numerical computation which a few individuals have possessed.

There are well attested cases of this power in virtue of which the mind reaches the result of difficult calculations by a species of Intuition—that is to say, without any consciousness of the process by which that result is made apparent to the Mind. This is not a proof that there is no process, but only that it is a process gone through as a machine goes through a process—that is, according to its own pre-adjusted laws of Motion. Perhaps, indeed, this process may not be different in kind from the process by which the average mind reaches the most elementary of arithmetical truths. The product of one and one, or of two and two, may be self-evident to all of us only in the same way in which the product of a long series of figures may be self-evident to minds with an abnormal gift of the arithmetical faculty. Thus the distinction breaks down between self-evident truths and truths which are not self-evident. A truth may be self-evident to one mind which is not self-evident to another, but may require, on the contrary, a laborious process of verification. And does not this again, lead us to see how entirely dependent are the phenomena of Mind upon the power of special Faculties, and how this power is itself dependent on the Adjustments of Organisation? In the world of Physics, we know that we are surrounded by movements which never make themselves sensible to us—pulsations which excite in our eyes no sense of light

—and others which excite in our ears no sense of sound, —and all this for want of adjusted organs. And so it would seem as if the Mind of Man were an Instrument attuned only to a certain range of knowledge, but as if within that range it were capable of finer and finer adjustments to the harmonies of Truth. These cannot make themselves heard where there is no organ to catch the sound. Nor could that organ translate them into Thought—into that conscious apprehension of which an Idea essentially consists,—had it not its own pre-adjusted relation to the Verities of the World.

It must be remembered, however, that in the discussion of such questions as to the Origin of our Ideas, there has been a great want of definition in the use of terms. Are fear, and love, and hatred, and anger, and jealousy, and remorse, and joy,—are these "ideas," or are they only conditions or powers of mind? If by Ideas we mean those imaginings which, as the very word implies, involve "images" of external things, it is certain that contact with external impressions, and, in this sense, Experience, is essential to the formation of them. But if by Ideas we mean the elementary passions, or if we mean even those peculiarities of thought—those special tendencies of mind which lead us to view things in some particular light rather than in others, and which constitute the essential distinction between the ideas of different men—if, in short, we include in the term anything which belongs to the Thinking Faculty itself, or anything of the method according to which it works up the raw material of Thought—then it is equally certain that Ideas in this sense are born with all of us, and that Imitation, and Experience, and Association, do but pour their material into moulds already cast for their reception.

But in reality here, as in many other questions, the rival disputants have each had some portion of the truth. They have been both right and both wrong. An Idea is not a simple, but a composite thing. It has not one origin, but a plurality of origins. An Idea is, as it were, a fabric of which the threads come from the spinner, and the weaving from the loom. Or it is, as it were, an organic growth, of which the materials are supplied from the external world, and the structure from the world within. There are many elements in every Idea which come, and can only come, from without. There are other elements, and among them the Formative Power, which come, and only can come, from within. The Mind stands in pre-established relations to the things around it—bound to them by the infinite adjustments which may be called External Correlations of Growth. Out of these relations it is not itself, nor do its powers possess the materials whereon to work. We cannot conceive a mind having no points of contact with the external world. From that world must come all the exciting causes of Thought and of Emotion. But the form into which these are cast—the tissue into which these are woven—the force by which Ideas become a Power—all, in short, that constitutes Thought as distinguished from the things about which we think—all this comes from, and belongs to, the Mind itself.

Among the lower animals, young ones, taken from the litter or the nest, and brought up under conditions wholly removed from the teaching of their parents, whether by imitation or otherwise, will reproduce exactly all those habits of their race which belong to their natural modes of life. Many of these habits, perhaps it may be safely said all of them, imply Ideas—that is to say, they imply instincts; and instincts are in the nature of ideas—that is to say, they belong to the phenomena of Mind. And of this there is another indication in a fact which at first sight may seem trivial or irrelevant. It has been often said that one great difficulty in reasoning on this subject, is the inaccessibility to observation of the mental condition of all infant creatures. But even if this were more true than it really is, there are some creatures, not low in the scale of creation, of which it may be said that, comparatively, they have no infancy at all. These are the Gallinaceous Birds in general, and some Species in particular. They come forth from the egg perfect miniatures of their parents, and with minds as fully equipped with parental instincts as their bodies are provided with feather or their wings with quills. Antecedent to all experience of injury, they exhibit fear, and not only fear, but fear of the proper objects. They will flee when they see a hawk, and they will carefully avoid a stinging insect. In Europe the young of the Woodgrouse or Gelinotte are able to fly from the moment they break the shell. In Australia, and the great group of islands which connect Australia with the Asiatic continent, there is a still more curious example of the same fact. There is a Family of Birds (*Megapodidæ*) of which the young are hatched, not by the incubation of the parents, but by the heat of fermentation generated in earthen mounds, scraped together for the purpose. From the moment the young are hatched they feed themselves, and run, and fly, and roost on trees, as if the world on which they have just opened their eyes had been long familiar. It is said, indeed, that the Parent Bird watches the Hatching Mound, and is ready to escort the chicks upon their first appearance in the surrounding scrub. But the recognition of the Parent by the young, and the answer to her call, are the most remarkable of all among these proofs of intuitive ideas. "As a moth emerges from a

Chrysalis, dries its wings, and flies away, so the young Telegallus, when it leaves the egg, is sufficiently perfect to be able to act independently.[1] Nor is this all; the curious instinct by which the Bird prepares an artificial Incubator for its young is an instinct born with it—an Innate Idea expressing itself in congenital habits of body. The chick of another Species of this singular family of Birds, the Megapode, was found in confinement to be incessantly scraping up sand and gravel into heaps, and the rapidity and power with which it effected this operation is described with astonishment by its captor.

These may seem far-fetched illustrations, and of slight value in so dark a subject; but let us remember that there are no solitary facts in Nature. There are indeed extreme cases,—extreme examples of universal laws,—that is to say, of laws whose operation is ordinarily restrained within narrower limits. But there is no fact standing really alone—not one which is not bound to the whole Order of Nature by deep analogies. That any creatures should be ushered into life so completely organised and furnished as the young of the Gallinaceous Birds and of the Megapodes, is a fact of immense significance in the phenomena of Organic Life.[2]

[1] Gould's "Birds of Australia."

[2] See Note E.

In Man analogous facts appear, modified by his infinitely wider range of character, and the infinite degrees in which the different elements of Mind are capable of being mixed in him. But although these conditions greatly complicate the result, the general phenomena are the same. Orphans, who have never had any opportunities of acquiring, by imitation, the peculiarities of their parents, will often, nevertheless, reproduce these peculiarities with curious exactness. This is a familiar fact, and how much this fact implies! Even when the inheritance is merely some congenital habit of body, or some trick of manner, it may, probably, imply some resemblance deeper than appears. For the Body and the Mind are in such close relationship, that congenital habits of Body are sure to be connected with congenital habits of Mind. But the inheritance is very often, so far as we can see, purely mental. How often do we recognise the tone, character, and the very turn of thought of dead friends, in the conversation and conduct of their children! The innate tendency to look at things in the same point of view, is evidenced in the reproduction of the same mental combinations, of the same images, of the same opinions, in short, of the same ideas. Cases, more remarkable than others of this kind, attract our attention, and we at once recognise ideas as innate which are so obviously determined by the forces of hereditary transmission. But we forget how often these laws of inheritance must be working invisibly where they never break ground upon the surface. And thus it is brought home to us how the Mind may be subject to laws of which it is unconscious—how its whole habit of thought, and the aspect in which different questions present themselves to its apprehension, are in a great measure determined by the mysterious forces of congenital constitution. And what is true in one measure of the individual mind, is true, also, in other measures, of whole families and of races of Men.

But the laws of Material Organisation are not the only laws to which Mind is subject. Obscure as these laws are, there are others which are obscurer still. What we cannot see in detail, we can see in the gross. What we cannot recognise in ourselves, we are able to recognise in others. We can see that the actions and opinions of men, which are the phenomena of Mind, do range themselves in an observed Order, upon which Order we can found, even as we do in the material world, very safe conclusions as to the phenomena which will follow upon definite conditions. And when we go back to former generations—to the history of nations, and the progress of the human race—we can detect still more clearly an orderly progress of events. In that order the operation of great general causes becomes at once apparent. On the recognition of such causes the Philosophy of History depends; and upon that recognition depends not less the possibility of applying to the exigencies of our own time, and of our own society, a wise and successful legislation.

But what are these causes, and what is the nature of those "laws" to which voluntary agents are unconsciously obedient? Is man's Voluntary agency a delusion, or is it, on the contrary, just what we feel it to be, and is it only from misconception of its nature that we puzzle over its relation to Law? We speak, and speak truly, of our Wills being free; but free from what? It seems to be forgotten that Freedom is not an absolute but a relative term. There is no such thing existing as absolute freedom—that is to say, there is nothing existing in the world, or possible even in thought, which is absolutely Alone—entirely free from inseparable relationship to some other thing or things. Freedom, therefore, is only intelligible as meaning the being free from some particular kind of restraint or of inducement to which other beings are subject. From what, then, is it that our Wills are free? Are they free from the influence of motives? Certainly not. And what are motives? A motive is that which moves, or tends to move, the mind in a particular direction. Like all other words which are used to describe the phenomena of Mind, it is taken from the language applicable to material things, and suggests the analogies which exist between them. It belongs to the profound but unconscious metaphysics of Human Speech. That which moves the Mind in a particular direction is best conceived of as something which exerts a force upon it, and the aggregate of such forces may, in a general sense, be called the laws which determine human action and opinions.

But here we come upon the great difficulty which besets every attempt to reduce to system the laws or forces which operate on the Mind of Man. It is the immense, the almost boundless variety and number of them. This variety corresponds with the variety of powers with which his Mind is gifted. For pre-established relations are necessary to the effect of every force, whether in the material or in the moral world. Special forces operate upon special forms of matter, and except upon these, they exert no action whatever. For no force can operate except where there are pre-established relations between its energies and the things upon which its energies are to work. The Polar Force of magnetism acts on different metals in different degrees, and there is a large class of substances which are almost insensible to its power. In like manner there are a thousand things that exercise an attractive power on the mind of a civilised man, which would exercise no power whatever upon the mind of a savage. And in this lies the only difference between subjection to Law under which the lower animals are placed, and the subjection to Law which is equally the condition of Mankind. Free Will, in the only sense in which this expression is intelligible, has been erroneously represented as the peculiar prerogative of Man. But the Will of the lower animals is, within their narrow spheres of action, as free as ours. A man is not more free to go to the right hand or to the left than the Eagle, or the Wren, or the Mole, or the Bat. The only difference is, that the Will of the lower animals is acted upon by fewer and simpler motives. And the lower the organisation of the animal, the fewer and simpler these motives are. Hence it is that the conduct and choice of animals—that is, the decision of their Will under given conditions—can be predicted with almost perfect certainty. Their faculties, few in number and limited in range, are open only to the small number of forces which are related to them; and in the absence of higher faculties accessible to other motives, these few attractions exert a determining effect upon their Will.[1]

Accordingly we may see that, in proportion as there is an approach among the lower animals to the higher faculties of Mind, there is, in corresponding proportion, a

[1] See Note F.

difficulty in predicting their conduct. Perhaps the best illustration of this is a very homely one—it is the effect of baits and traps. Some animals can be trapped and caught with perfect certainty; whilst there are others upon which the motive presented by a bait is counteracted by the stronger motive of caution against danger, when a higher degree of intelligence enables the animal to detect its presence. Yet the Will of the cunning animal is not more free than the Will of the stupid animal,—nor is the Will of the stupid animal more subject to Law than the Will of the cunning one. The Will of the young Rat, which yields to the temptation of a bait, and is caught, is not more subject to Law than the Will of the old Rat, who suspects stratagem, resists the temptation and escapes. They are both subject to Law in precisely the same sense and in precisely the same degree—that is to say, their actions are alike determined by the forces to which their faculties are accessible. Where these are few and simple, the resulting action is simple also; where these are many and complicated, the resulting action has a corresponding variety. Thus the conduct of animals is less capable of being predicted in proportion as it is difficult or impossible to foresee the nature or number of the motive forces which are brought to bear upon the Will. Man's Will is free in the same sense, and in the same sense only. It is subject to Law in the same sense, and in the same sense alone. That is to say, it is subject to the influence of motives, and it can only choose among those which are presented to it, or which the mind has been given the power of presenting to itself.[1]

But in this last power we touch the secret of that boundless difference which separates Man from the highest of the animals below him. There is such a gulf between the faculties of his mind and those of the lower animals, that the forces acting on the human spirit become, by comparison, innumerable, and involve motives belonging to a wholly different class and order. He is exposed, indeed, to the lower motives in common with the beasts. But there are others which operate largely upon him which never can and never do operate upon them. Foremost among these are the motives which Man has the power of bringing to bear upon himself, arising out of his power of forming Abstract Ideas, out of his possession of Beliefs, and, above all, out of his Sense of Right and Wrong. So strong are these motives that they are able constantly to overpower, and sometimes almost to destroy, the forces which are related to his lower faculties. Again, among the motives which operate upon him, Man has a selecting power. He can, as it were, stand out from among them,—look down from

[1] See Note G.

above them, — compare them among each other, and bring them to the test of Conscience. Nay more, he can reason on his own character as he can on the character of another Being,—estimating his own weakness with reference to this and the other motive, as he is conscious how each may be likely to tell upon him. When he knows that any given motive will be too strong for him, if he allow himself to think of it, he can shut it out from his mind by "keeping the door of his thoughts." He can, and he often does, refuse the thing he sees, and hold by another thing which he cannot see. He may, and he often does, choose the Invisible in preference to the Visible. He may, and he often does, walk by Faith and not by Sight. It is true that in doing this he must be impelled by something which is itself only another motive, and so it is true that our Wills can never be free from motives, and in this sense can never be free from "Law." But this is only saying that we can never be free from the relations pre-established between the structure of our minds, and the system of things in which they are formed to move. From these, it is true indeed, that we never can be free. But as a matter of fact, we know that these relations do not involve compulsion. It is from compulsion that our Wills are free, and from nothing else; and for this freedom we have the only evidence we can ever have for any ultimate truth respecting the powers of Mind—the evidence of Consciousness—that is, the evidence of observation turned in upon ourselves.

The discussions of many centuries seem to have resulted, at last, in some real progress upon this vexed question of Necessity and Free-will. That progress lies mainly in a clearer definition of terms. The most eminent living philosopher who represents the doctrine, commonly called the Doctrine of Necessity, repudiates that name as incorrect, expressly on the ground that the word Necessity, as commonly applied, signifies compulsion. Undoubtedly it does; and if this meaning be repudiated, then the word is not used in its ordinary and legitimate sense. This, indeed, Mr. Mill confesses, whilst yet he casts upon his opponents the blame of a misunderstanding, which assuredly lies with those who do not employ ordinary words in the ordinary signification. "The truth is," he says, "that the assailants of the doctrine (of Necessity) cannot do without the associations engendered by the double meaning of the word Necessity, which in this application signifies only invariability, *but, in its common employment, compulsion.*"[1] He believes, therefore, in Necessity only in the sense of Invariability. But if the doctrine which Mr. Mill favours has suffered from one ambiguity, it seeks to shelter itself

[1] "Examination of Sir W. Hamilton's Philosophy," by J. S. Mill, p. 492, note.

under the protection of another ambiguity much more deceptive. If there is a double meaning in the word Necessity which has exposed the Necessitarian doctrine to unjust objections, it is equally true that there is a double meaning in the word Invariability which lends to that doctrine an undue advantage. Invariability can be predicated of mental action in this vague general sense —that all the movements of Mind must invariably arise from some motive. But this is a kind of "Invariability" which admits of any amount of variation. For, as in the language of this philosophy, Necessity does not mean compulsion, so by Invariability, as applied to the phenomena of Mind, nothing more is meant than that, in respect to mental action, there is an "abstract possibility of its being foreseen." "If," says Mr. Mill, "necessity means more than this abstract possibility of being foreseen; if it means any mysterious compulsion, apart from simple invariability of sequence, I deny it as strenuously as any one."[1]

But now let us insist, as in such subjects we are bound to do, on still clearer definitions. We shall find, in the first place, that the "abstract possibility" of foreseeing mental action depends on nothing less than such absolute knowledge of character and of motive

1 "Examination of Sir W. Hamilton's Philosophy," p. 517. See Note H.

as can belong to God alone. We shall then find, in the second place, that this favourite phrase, "invariability of sequence," is as ambiguous as others of the same class. It does not mean that any particular sequences are invariable, but only that there must always be *some* sequence—that it is invariably true that everything which happens has proceeded from *something* as a cause, and leads to *something* as a consequence. But this is a proposition which evidently, when reduced to its true dimensions, has no adverse bearing whatever on the doctrine of Free Will. The "abstract" possibility of foreseeing mental action depends on these two propositions: first, that where *all* the conditions of that action are constant, the resulting action will be constant also; and, secondly, that absolute and perfect knowledge of the whole of those conditions would carry with it sure foreknowledge also of the choice to which they lead. But surely this is not only true, but something very like a truism.[1] There is nothing to object to

[1] Mr. Mansel, following other philosophers on this point, reduces the modified doctrine of Necessity to this identical proposition, "that the prevailing motive prevails." Mr. Mill's reply is altogether unsatisfactory.—*Examination of Sir W. Hamilton's Philosophy*, pp. 518, 519.

I cannot help adding here—although the observation has reference to another subject—that Mr. Mill appears to me to have exposed with great force and clearness the verbal fallacies involved in Mr. Mansel's work on the "Limits of Religious Thought," and espe-

or deny in the doctrine, that if we knew *everything* that determines the conduct of a man, we should be able to know what that conduct will be. That is to say, *if* we knew *all* the motives which are brought by external agencies to bear upon his mind, and *if* we knew all the other motives which that mind evolves out of its own powers, and out of previously acquired materials, to bear upon itself; and *if* we knew the character and disposition of that mind so perfectly as to estimate exactly the weight it will allow to all the different motives operating upon it,—*then* we should be able to predict with certainty the resulting course of conduct.

This is true, not only as an abstract conception, but as a matter of experience in the little way towards perfect knowledge along which we can ever travel. We can predict conduct with almost perfect certainty when we know character with an equal measure of assurance, and when we know the influences to which that character will be exposed. In proportion as we are sure of character, in the same proportion we are sure of conduct. Yet we never think of the Will being the less free, because we can predict its course. What we know in such cases is simply the use which, under given conditions, will be

cially in the use he makes of such forms of expression as "The Absolute," "The Infinite," &c.—See the chapter (vii.) on "The Philosophy of the Conditioned as applied by Mr. Mansel to Religion," in the same work.

made of freedom. There is no certainty in the world of Physics more absolute than some certainties in the world of Mind. We know that a humane man will not do a uselessly cruel action. We know that an honourable man will not do a base action. And if in such cases we are deceived in the result, we know that it is because we were ignorant of some weakness or of some corruption; that is to say, we were ignorant of some elements of character. But we never doubt that, if those had been known, we could have foreseen the resulting lapse. Perfect knowledge must therefore be perfect foreknowledge. To know the present perfectly, is to know the future certainly. To know all that is, is to know all that will be. To know the heart of Man completely, is to know his conduct completely also; for "out of the heart are the issues of life." So far from this conclusion being dangerous or hostile to any part of the Christian system, it is a conclusion which enables us, in a dim way, not merely to hold as a Belief, but to see as a necessary truth, that there can be no chance in this world,—and how it is, and must be, that to the All-seeing and All-knowing the Future is as open as the Present and the Past. But none of these ideas involve the idea of compulsion; and the absence of compulsion is all that can be meant by Freedom.

And as by Freedom, we do not mean freedom from

motives, so neither do we mean that any of the phenomena of Mind, any more than any of the phenomena of Matter, can arise without "an antecedent." In this sense there is no contradiction between the doctrine of Free Will and the amended doctrine of Necessity. Man is subject to the law of Causation in this sense, "that his volitions are not self-caused, but determined by spiritual antecedents in such sort that when the antecedents are the same, the volitions will always be the same."[1] But this word "antecedent" is one of the many vague words in which metaphysicians delight. The highest antecedents which we can ever trace as determining conduct, are to be found in the constitution of mind itself. Love is an antecedent, so is Reverence, so is Gratitude, so is the Hunger after Knowledge, so is the Desire of Truth. So also is the action of other Spirits upon our own. Higher than these—further up the chain of Cause and Effect—we cannot go. And yet we need not conceive of these as "Final Causes," nor does the doctrine of our Free Will assign to the human Mind any self-originating power. Man has nothing which he did not receive. Such freedom as his Will possesses has been given to him, and given him, too, as we have dimly seen, by the employment and by the device of means. It is a power belonging

[1] "Mill on Hamilton," pp. 492, 493.

to his structure, and derived from Him by whom that structure has been devised.

"Our Wills are ours, we know not how."

The power which in health we possess of preferring one motive to all others, whilst yet the influence of those others may be strongly felt, is a power which like every other, must have its own "antecedent"—that is to say, its own cause, and its own purpose. But these are to be found in the Adjustment from which the power arises,—in the Mind by which that adjustment has been contrived, and in the Purposes which it reveals. The freedom of Man's Will is not more mysterious, when it is exerted in directing the Mind to one motive, and averting it from another, than when it is exerted in turning the Body to the right hand rather than to the left.[1]

The difficulty of reconciling, in one clear Order of Thought, the idea of the Freedom of our own Will with the idea of Causation, is not really so great a diffi-

[1] The whole of this passage on Necessity and Free Will has been severely criticised in an article in the *Dublin Review* for April 1867, as involving a practical abandonment of the very doctrine which I profess to defend. The argument there maintained seems to me altogether erroneous; and I have seen no reason to alter the text in any material point. The subject, however, is so important in itself, and so interesting as regards the history of Philosophy, that I have thought it right to deal with it in a separate note (F) already referred to.

culty as the use of ambitious and ambiguous language has made it appear to be. There are two sentences in Mr. J. S. Mill's work, on the Philosophy of Comte, which afford the best possible illustration both of the true doctrine on the relation in which Will stands to Law, and of the false doctrine into which it may be merged by the ambiguous use of words. In one passage Mr. Mill defines the Positive as distinguished from the Theological Mode of Thought to be—"that all phenomena, without exception, are governed by invariable laws, *with which no volitions either natural or supernatural interfere.*"[1] It is at least satisfactory to find in this sentence so clear an avowal that the idea of free Divine Volition in the region of the Supernatural, and the idea of free Human Volition in the region of the Natural, stand on the same ground, are exposed to the same intellectual difficulties, and are both equally denied by the new Philosophy. But as a definition of the Positive mode of thought it stands in curious contrast with another passage of the same work, in which Mr. Mill says that "the Theological mode of explaining phenomena was once universal, *with the exception, doubtless, of the familiar facts which being even then seen to be controllable by human Will belonged already to the Positive Mode of Thought.*"[2]

1 "Aug. Comte and Positivism," p. 12. 2 Ibid. pp. 31, 32.

These two sentences involve, on the face of them, contradictory positions. The one affirms that no volitions can interfere with the laws which govern phenomena, and that the recognition of this is the very essence of the Positive Philosophy. The other affirms that the Positive Mode of Thought is involved in the very idea of facts being controllable by human Will.

It is not, perhaps, very important to ask which of these two sentences gives the most accurate description of the Positive Philosophy; but it is of much importance to ask which of these two positions is nearest to the truth? Beyond all doubt, it is the last. If the Positive Philosophy were content with the assertion that the power of Will over facts depends on the invariability of Laws—that is, on the constancy of Natural Forces—it would be sound enough. And so, the second of the two sentences I have quoted sets forth the central idea of that Philosophy in its most favourable light. But in the first of those two sentences we have a concentration of all that is erroneous in Positivism, and at the same time a typical example of the ambiguities and obscurities of language on which the fallacies of that Philosophy depend. There is hardly a single word in that sentence which is not ambiguously used. "Phenomena" and "facts," "govern" and "control," and "interfere with," are all used in ambiguous senses; whilst, as usual, the words "Law" and "Invariable," are used not only ambiguously, but unintelligibly. In order to test these ambiguities we have only to compare the two sentences together. "Phenomena" in the one sentence seems to correspond with "facts" in the other. Yet, we have this result,—that "phenomena" are governed by Invariable Law, whilst "facts" are controllable by human Will. It would appear, then, that the "phenomena" which are governed by Law cannot be the same with the "facts," which are controllable by Will:—or else, if they be the same, then there must be some essential distinction between "controlling" and "governing." What is this distinction? It is not defined, or even suggested. Then, again, if no volitions can "interfere with" Laws, how can volitions "control" facts? If Will controls facts, and yet can't "interfere with" Laws, how is the control over facts exercised? What is the relation between the Laws which no volitions can "interfere with," and the "facts" which volitions do actually "control?" Can Will control facts, which again are governed by laws, (in some sense or other) either by interfering with those laws, or controlling them?

If it were possible to get any definite meaning out of this confusion of words, perhaps it might be said that Will can "control" Law, but cannot "interfere with" it. There is at least a glimmering of the truth in this. But no man could gather from those two sentences of Mr. Mill what the truth is, although, after all, the truth is plain enough, if only some care be taken to confine definite words to some sort of definite meaning. If by Laws are meant the elementary Forces of Nature, and if by "interfering" with them is meant any power of altering their own essential energies—then it is true that no volitions of ours can interfere with them. But then it cannot be too often repeated that, in this sense, phenomena are NOT governed by Invariable Laws; because phenomena are never the result of individual Forces, but are always the result of the conditions under which several Forces are combined, and these conditions are always variable. If, again, "interference" means or includes the power of setting Natural Forces (Laws) to work under new conditions, then it is the reverse of truth to affirm that they cannot be "interfered" with. Man controls facts only because (in this sense) he can, and he does, interfere with Laws. His volitions can, and do, govern those combinations of Force which are the immediate cause of all phenomena.

There is no fault in philosophical discussion more pestilent than that of using common words in some technical or artificial sense, without any warning to the reader, (often apparently without any consciousness on the part of the writer,) that ideas fundamentally involved, in the ordinary use of the word, are eliminated and set aside. We have seen one instance of this in the word "necessity," emptied of its meaning of compulsion. We have another example in the use made of such words as "changeable," and others of a like kind. Thus Mr. Mill[1] quotes, with approbation, a remark of Comte, that "our power of foreseeing phenomena, and our power of controlling them, are the two things which destroy the belief of their being governed by changeable Wills." All through this sentence there run the same confusions which have been pointed out in the two sentences already quoted. But there is, in addition, another confusion which has a special bearing on the subject of this chapter. Phenomena which can be controlled are phenomena which can be changed. There is no other meaning in the words. The assertion, therefore, is, that the changeability of phenomena through human agency is a fact which must destroy our belief in the changeability of the human Will itself. The sentence thus rendered is, of course, either pure nonsense, or else must be dependent for a rational sense upon some artificial meaning being attached to the word "changeable." A Will under the guidance of some settled principle—that is to say, following habitually some prevailing motives—

1 "Auguste Comte and Positivism," p. 48.

might, by a certain licence of language, be called an unchangeable Will. But this has nothing to do with that kind of changeability which can alone concern the power of altering and controlling material phenomena. Stability of character, whether moral or purely intellectual, is not only compatible with a variable Will, but it is inseparably connected with it. No man can pursue one rule of conduct under changing conditions unless he himself retains his own capacities of change. He cannot control phenomena without changing them, and he cannot change phenomena without changing his own course of action; and a change in the course of action is a change in the course of Will.

That which is really at the bottom of all this ambiguity of language, is a constant endeavour to get rid altogether of an essential element in the very idea of Will,—to reduce it to something different from that which we all know and feel it to be. The word Will is indeed retained in the Positive vocabulary, but some other word is generally inserted before it, to prejudice the common understanding of it, or to impart some element of meaning which can with more plausibility be denounced. Thus the Will which is denied in Nature is often described as an "arbitrary" Will or a "capricious" Will. But surely these qualifying epithets do but add to the confusion. It is true, indeed, that the Will we see in

Nature is not a capricious Will. But this is not the question. The question is, whether there is, or is not, such a thing possible as caprice in Will. If there be such a thing as caprice, then the existence of it, and the power of it "to control phenomena," cannot be denied. If there be no such thing, then "capricious" is of no meaning as an epithet applied to Will. Caprice implies not only changeableness, but, so to speak, a double degree of changeableness—a changeableness which has no rule or reason in its shiftings. It is a fact that there are human Wills of this character, and the mischief they have done in the world arises from the power they possess, in common with all other Wills, of changing phenomena after their own unreasonable nature. The truth is, that if the human Will can be described as unchangeable, then there is no such thing as changeability even conceivable in thought. There is no contrast so absolute between any two different forms of Matter, as there is between two different states of the same Mind. There is no transition in Nature from one physical condition to another so absolute or so radical as the transition to which human character is subject when it passes under the power of new convictions. There is no change like the change from hatred to affection, from vice to virtue, from evil to good. And this change in Mind is the efficient cause of a whole cycle of other changes among the phenomena which the human Will can and does alter, regulate, and control.

There is, then, not much real difficulty after all in disengaging the great facts of our own Free Will from the verbal confusions of the Positive Philosophy. Nor will the same methods of solution fail us when we apply them to the further question,—How far, and in what sense, are our own volitions themselves subject to law—that is, to the influence of Adjusted Forces? For as one great consequence of the Reign of Law over material things is the necessity of resorting to the use of appropriate means for the accomplishment of Purpose, so does the same necessity arise out of the same conditions among the phenomena of Mind. If we wish to operate upon human action, we must go to work by presenting to the Will some motive tending to produce the action we desire. Above all, if we seek to operate not merely on individual actions, but upon that which mainly determines conduct, viz. human character, we must direct our efforts to place that character under outward conditions which we know to have a favourable effect upon it. In the material world we should be powerless to control any event if we did not know it to be subject to laws—that is, to Forces which, though not liable to change in essence, are subject to endless change in combination and in use. The same impotency would affect us, if in the moral world also definite conditions had not always an invariable tendency to produce certain definite results. It is a mere confusion of thought and of language which confounds the "invariability" of "Laws," either moral or material, with the denial of the power of Will to vary, alter, and modify in infinite degrees the course of things. It is the fixedness of all Forces in one sense which constitutes their infinite pliability in another. It is the unchanging relation which they bear to those mental faculties by which we discover them and recognise them, that renders them capable of becoming the supple instruments of those other faculties of Will, of Reason, and of Contrivance by which we can work them for altered and better purposes.

CHAPTER VII.

LAW IN POLITICS.

AT first sight it may be thought that the means by which we can operate on the Wills of individual men, and of communities of men, are contained within a narrow compass, and are such as to be all, if not within easy reach, at least within easy recognition. And it is true that some methods of operating on the minds of men we do know instinctively, just as in the material world we know by the first rudiments of intelligence how to accomplish a few physical results. But experience and observation teach us, although they teach us very slowly, that direct appeals to the reason, or direct appeals to the feelings of men, are entirely useless, when those faculties have not been placed under conditions favourable to their exercise in a right direction. And as in the material world, the knowledge we have acquired of the powers of Nature, and of the methods of turning them to use, has been slowly gained in the lapse of ages, and as all we discover does but reveal how much we have yet to know; so in the immense world of the Mind and

Character of Man, our knowledge of the methods by which it may be well and wisely governed, has advanced only by slow degrees. There is a boundless field of discovery still open to those who investigate the laws which govern the development of our nature. When we look at the high degrees of excellence which that nature so often attains under favourable conditions for the growth and exercise of its better powers, and when we contrast this with its stunted and distorted growth as exhibited among large portions of Mankind, it becomes a question of deep and endless interest to know how far these conditions are subject to the control of Will through the use of means. If such means can ever be devised, it must be by knowledge, first of the elementary forces which have a constant operation on Human Character, and secondly by contrivance in so combining them as to make them operate in the direction we desire. And it is in this search that we discover the intimate blending and inseparable connexion between mental and material laws—that is, between the forces which operate on the material frame and the forces which operate on the Mind and Character of Man.

And here we come on a great subject—the function of Human Law as distinguished from Natural Law. Just as the Will of the individual can operate upon itself by the use of means, some of which are known instinctively, whilst others are found out by reason; so can the collective Will of Society operate upon the conduct of its members in two ways—first, directly by authority; and secondly, indirectly by altering the conditions out of which the most powerful motives spring. This last is a principle of government, which has been distinctly recognised only in modern times, and which admits of applications not yet foreseen. The idea of founding Human Law upon the Laws of Nature, is an idea which, though sometimes instinctively acted upon, was never systematically entertained in the ancient world. Indeed, the true conception of Natural Law is one founded on the progress of physical investigation, and growing out of the habits of scientific thought. It was long before Man came to apprehend the prevalence of Law in the phenomena of Matter; and it was still longer before he could even entertain the notion of Natural Law as applicable to himself. The ancient lawgivers were always aiming at standards of Political Society, framed according to some abstract notions of their own as to how things ought to be, rather than upon any attempt to investigate the constitution of human nature as it actually is. It was a mistake in the science of Politics analogous to that which Bacon complained of so bitterly in the science of Physics. Men were always trying to evolve out of their own minds knowledge which could only be acquired by patient inquiry into facts. How worse than useless this method is, received an illustration in ancient philosophy still more striking than in ancient legislation. Fortunately for mankind, no actual legislators have ever been quite so foolish as some philosophers. Perhaps, all things considered, the most odious conceptions of Human Society which the world has ever seen, were the conceptions of an intellect certainly among the loftiest which has ever exercised its powers in speculative thought. Plato's Republic is an Ideal State, founded on abstract conceptions of the mind, and one of its leading ideas is the destruction of Family Life, and the annihilation of the family affections. And yet this result, odious and irrational as it is, was arrived at from reasoning which is not in itself odious, but which is false, chiefly because it takes no account of the facts of Nature. The welfare of the State was to be the one object of desire in every mind. All separate interests and affections were to be suppressed, and amongst these the very idea of special property in Wife or Child. The highest type of man was to be bred by the Republic as the highest type of dogs and horses is bred by an intelligent owner.[1] Such are the humiliating results of abstract reasoning,

[1] "The breeding is regulated, like that of noble horses or dogs, by an intelligent proprietor."—Grote's "Plato," vol. iii. p. 203.

pursued in ignorance of the great Law, that no purpose can be attained in Nature except by legitimate use of the means which Nature has supplied. For as in the material world, all her Forces must be acknowledged and obeyed before they can be made to serve, so in the Realm of Mind there can be no success in attaining the highest moral ends until due honour has been assigned to those motives which arise out of the universal instincts of our race.

Accordingly it is remarkable that the system of ancient philosophy, which for so many ages continued to rule the thoughts of men—the philosophy of Aristotle—owes almost all the strength it has in Politics as in other matters, to occasional and almost unconscious resort to the true methods of scientific reasoning and investigation. Aristotle founds his adverse criticism on Plato, where it is most successful, upon the actual facts of what men, under specified conditions, naturally do, and think, and feel. From these facts he argues justly as to what they would do under the artificial restrictions of a theoretical philosophy. When, for example, he argues against communism, and in favour of private property, upon the ground of the watchfulness and attention which self-interest produces in the conduct of business,[1] and when he adds, "It is unspeakable how advantageous it is that

[1] μᾶλλον δ' ἐπιδώσουσιν ὡς πρὸς ἴδιον ἑκάστου προσεδρεύοντος.—"Aristot. Pol." Bk. ii. c. 5.

a man should think he has something which he may call his own, for it is by no means to no purpose that each person should have an affection for himself, *for that is natural*,"[1] he touches the very root idea of the modern science of Political Economy. He touches it, but he does not grasp it. It is a line of argument which is never consistently maintained; and though there are perpetual appeals to "nature"—to that which is "natural"—to that which nature teaches—no definite meaning can be attached to these expressions; and dogmas are laid down as "natural" which are purely abstract and metaphysical conceptions. Nature is called as a witness, and then the witness she gives is condemned and put out of court. Industry is occasionally praised, whilst the means and the motives to industry are systematically despised. The exercise of any mechanical employment, or the following of merchandise, is condemned in an Ideal Government as "ignoble and destructive to virtue."[2] A maritime situation is recommended, because of its convenience in enabling a city to receive from others produce which its own country

[1] ἔτι δὲ καὶ πρὸς ἡδονὴν ἀμύθητον ὅσον διαφέρει τὸ νομίζειν ἴδιόν τι· μὴ γὰρ οὐ μάτην τὴν πρὸς αὑτὸν αὐτὸς ἔχει φιλίαν ἕκαστος, ἀλλ' ἔστι τοῦτο φυσικόν.—Bk. ii. c. 5.

[2] οὔτε βάναυσον βίον οὔτ' ἀγοραῖον δεῖ ζῆν τοὺς πολίτας· ἀγεννὴς γὰρ ὁ τοιοῦτος βίος καὶ πρὸς ἀρετὴν ὑπεναντίος.—Bk. vii. c. 9. In Mr. Congreve's edition, Bk. iv. c. 9.

does not afford, and to export those necessaries of life of which it has more than plenty. This looks like a perception of the soundest maxims of Commerce. But in the next breath, the whole richness and blessing of Commerce, as an element of civilisation, is repudiated and destroyed by the stupid and selfish maxim that a city must traffic to supply its own wants only, and not the wants of others: "for those who make themselves into an open market for every one, do it for the sake of revenue; but if a State ought to have no part in this kind of gain, neither ought it to furnish such a mart." [1]

It is surely wonderful that such a mind as that of Aristotle should have supposed that it was either possible, or, if possible, desirable that the benefits of traffic should all be on one side; nor is it less wonderful that, with his hands, as it were, upon the spot, and touching with his very fingers the foundation-facts, he should yet have failed to feel and to seize the great secret of modern Political Science—the links of Natural Consequence in which the blessedness of Commerce lies. But all this comes of thinking that we can be wiser than Nature, and of failing to see that every natural instinct has its

[1] *αὐτῇ γὰρ ἐμπορικὴν, ἀλλ' οὐ τοῖς ἄλλοις δεῖ εἶναι τὴν πόλιν. οἱ δὲ παρέχοντες σφᾶς αὐτοὺς πᾶσιν ἀγορὰν προσόδου χάριν ταῦτα πράττουσιν· ἣν δὲ μὴ δεῖ πόλιν τοιαύτης μετέχειν πλεονεξίας, οὐδ' ἐμπόριον δεῖ κεκτῆσθαι τοιοῦτον.*—Bk. vii. c. 6.

own legitimate field of operation, within which we cannot do better than let it alone. It comes from the notion that we can arrive at that which ought to be, without taking any note of that which actually is.

The bondage under which all true Science lies to fact —the necessity of groping among the detail of little and common things—this is a hard lesson for the human Intellect to learn—conscious as that Intellect is of its own great powers—of its own high aims—of its own large capacities of intuitive understanding. But it is a lesson which must be learnt. There are no short cuts in Nature. Her results are always attained by Method. Her purposes are always worked out by Law. So must ours be. For our bodies and our spirits are both parts of the great Order of Nature; and our Wills can attain no end, and can accomplish no design, except through knowledge and through use of the appropriate and appointed means. Nor can those means be ascertained except by careful observation, and as careful reasoning. It is a hard thing to know all the forces which operate even on our own individual minds; and it is a much harder problem to understand the forces which arise out of the complicated conditions of human society. But the very idea of Natural Law as affecting mankind is founded on the possibility of tracing in human nature the existence and operation of forces which under given conditions do actually determine the course of human conduct in particular directions. Amongst these forces there are a certain number which are constant, or at least so constant that they may be calculated upon as certainly affecting the great majority of mankind. These are chiefly the motives which arise out of our physical constitution—the desires and affections which are common to the race. To follow these motives—to be actuated by them—is, therefore, natural. And yet to follow these motives exclusively, may, and generally does, lead to great evils, often to calamities, sometimes to destruction. How, then, can these motives be controlled? Only by appealing to other motives—to forces lying in the higher regions of the mind, and placed there like the forces of external Nature, to be at the disposal of the Intelligence and the Will.

Are, then, these higher motives not also natural?—are they above nature?—are they supernatural? It would really seem as if this were the idea involved in the distinction which is so vaguely drawn between that which is said to be natural and that which is said to be not natural—between Natural Law and Positive Institution. Yet Reason, and Conscience, and Fancy, and Imagination, and Belief, or whatever other faculties may direct, wisely or unwisely, the course of legislation, are all equally natural to Man. They are all as much parts of his mental constitution as the desires and instincts to which the term natural is usually confined. There is no extravagance of the individual Will—there is no folly of blind and irrational legislation which has not been the fruit of some part or another of Man's nature. I dwell on this only because it is important here, as in other cases, to attach a definite meaning to the words we use, and especially to a word which plays so important a part in the language both of Philosophy and of Politics.

It appears, then, that, as applied to human conduct, we mean by "natural" conduct that which men are prompted to pursue rather by instinct and impulse than by calculation of consequences and by reason. Human Laws, or Positive Institutions, as being the result of deliberation, stand contrasted with Natural Law in this sense, and in this sense alone. For as Reason and Reflection are natural to Man, and are as important parts of his nature as the highest of his instincts, so Laws founded on a right exercise of that Reason are Natural Laws in the best and highest sense of all. Laws, however, whether in this sense natural or not—that is, whether founded on a right or a wrong exercise of reason—are always intended to act as restraints on the actions of individuals, and to interfere with the motives by which their conduct would be otherwise determined. This restraint may be said to be artificial as opposed to the natural restraints of the individual reason; and this, perhaps, is the distinction most generally intended when the natural conduct of men is contrasted with their conduct under the control of Positive Institution. But as the motives which determine individual conduct are not always reasonable motives, so it is clear that what men naturally do is no sure test either of what they ought to do, or of what they ought to be allowed to do. It is their nature, under certain conditions, to do all that is bad and injurious to themselves and others. Hence it is the most difficult of all problems in the Science of Government to determine when, where, and how it is wise to interfere by the authority of Law with the motives which are usually called the natural motives of men. The question is no other than this: How far the abuse of those motives can be checked and resisted by that public authority whose duty and function it is to place itself above the influences which, in individual men, overpower the voice of reason and of conscience?

No more signal illustration has been ever given of the relation between Natural Law and Human Law—of the circumstances in which Natural Law may be trusted, and of those in which it absolutely requires to be controlled—than the illustration afforded by the history of Legislation in our own country within the present century. During that period two great discoveries have been made in the Science of Government: the one is the immense advantage of abolishing restrictions upon Trade; the other is the absolute necessity of imposing restrictions upon Labour. The rise, the growth, and the final acceptance of these two ideas as the basis of practical Legislation, is a history so curious, and having such close relation to the subject of this chapter, that I propose to deal with it somewhat in detail.

Since the dissolution of the Greek and Roman Commonwealths, no nation has acted on the one great error of all the ancient systems of political philosophy—that the natural desire of men for the accumulation of wealth is an evil to be dreaded and repressed. So far as this goes there is a sharp and striking contrast between the spirit of ancient and of modern policy. The great object of the ancient policy, says Dugald Stewart, "was to counteract the love of money and a taste for luxury by positive institutions, and to maintain in the great body of the people habits of frugality and a severity of manners. The decline of States is uniformly ascribed by philosophers and historians, both of Greece and Rome, to the influence of riches on national character; and the laws of Lycurgus, which, during a course of ages, banished the precious metals from Sparta, are proposed by many of them as the most perfect model of legislation devised by human wisdom. How opposite to this is the doctrine of modern politicians! Far from considering poverty as an advantage to a State, their great aim is to open new sources of national opulence, and to animate the activity of all classes of the people by a taste for the comforts and accommodations of life."[1] This is true, and has been true more or less of all the modern nations of the world. But although they never held the absurd doctrine that Nature was wrong when she taught men to desire wealth, they did hold the doctrine, hardly less mischievous, that Nature was incompetent to teach them how best to acquire it. It would be difficult to say whether the law of ancient Sparta, prohibiting gold from ever coming into the State, was worse than the law of modern Spain, which prohibited gold from ever being allowed to leave it. It is certain that the Spanish law was at least the more irrational of the two. If a State wishes to be poor, it is not absurd to prohibit the making of money. But if a State wishes to be rich, it is mere stupidity to prohibit the natural use of the medium of exchange. Yet this law of Spain is only an extreme example of the system and the theories which governed, until the other day, the legislation of all the nations of Europe, and which still largely prevails amongst them.

1 "Account of the Life and Writings of Adam Smith," by Dugald Stewart.—"Collected Works of Dugald Stewart," vol. x. p. 57.

It was no oratorical exaggeration, but a strict and literal description of the truth, when Mr. Gladstone said[1] of the old commercial policy that it was "a system of robbing and plundering ourselves." And how was it so? What was the essence of its error? These questions are best answered by another. What was the central idea of the new system which has superseded the old one? The essential idea of these new opinions cannot be better given than in the words of Dugald Stewart: "The great and leading object of Adam Smith's speculations is to illustrate the provision made by Nature in the principles of the human mind, and in the circumstances of man's external situation, for a gradual progressive augmentation in the means of national wealth; and to demonstrate that the most effectual plan for advancing a people to greatness is to maintain that order of things which Nature has pointed out; by allowing every man, as long as he observes the rules of justice, to pursue his own interest in his own way, and to bring both his industry and his capital into the freest competition with those of his fellow-citizens."[2]

Adam Smith found Positive Institutions regulating and restricting natural human action in two different directions. There were laws restricting free interchange in

In his Speech at Glasgow, Oct. 1865.

Account, see p. 72.

the products of labour: and there were other laws restricting the free employment of labour itself. He denounced both. Labour was deprived of its natural freedom by laws forbidding men from working at any skilled labour, unless they had served an apprenticeship of a specified time. It was also deprived of its natural freedom by monopolies, which prevented men from working at any trade within certain localities, unless allowed to do so by those who had the exclusive privilege. The first mode of restriction prevented labour from passing freely from one employment to another, even in the same place. The second mode of restriction prevented labour passing freely from place to place, even in the same trade. Both of these restrictions were as mischievous, and as destructive of their own object, as restrictions in the free interchange of goods. They both depended on the same vicious principle of attempting to obtain by Legislation results which would be more surely attained by allowing every man to sell his goods or his labour when, where, and how he pleased. The labour of a poor man was his capital. He had a natural right to employ it as he liked. And as for protecting the community from bad or imperfect work, *that* would be best secured by unrestricted competition. The natural instincts and respective interests of producers and consumers would secure mutual adaptation. Perfect freedom of exchange in goods, the products of labour, and perfect freedom in the application of labour itself—this was the rule to follow. Natural Law was the best regulator of both. Such were the doctrines of Adam Smith, then new in the world.

It is not a little remarkable that, during the same years in which Adam Smith was working out his memorable Inquiry, other minds, working in a very different department of human thought, were preparing events which were to bring to a speedy test how far these doctrines of Natural Law were true absolutely, or true only under limitations, which he did not foresee. When Adam Smith was lecturing with applause in Glasgow from the chair of Moral Philosophy, James Watt was selling mathematical instruments in an obscure shop within the precincts of the same University. It may seem as if no two departments of human thought are more widely separated than those in which these two men were working. One was a region purely mental. The other was a region purely physical. The one had reference to the Laws of Matter. The other had reference to the Laws of Mind. Yet the work of James Watt and the work of Adam Smith were inseparably connected, not only as involving analogous methods of investigation, but as showing in their result the blending and co-operation of mental and material laws.

It was the labour of Watt to reduce to obedience, under the power of Mind, one of the most tremendous Forces of Nature, and this he did through many years of curious inquiry, and of laborious contrivance. He found only a rude and imperfect mechanism through which this great Force had been misdirected and dissipated and lost. He collected it in fitter vessels; he led it into smoother channels; he opened for it doors of passage, through which the rushing of its escape did for him what he wanted it to do. Other forces, which before had conspired against it, were so guided as to work along with it, not only in perfect harmony, but in close alliance. He made, in short, its invariable energies subject to the variable conditions of Adjustment. And so, he governed it and controlled it, and handed it over to the Human Family as the servant of their Will for ever.

The work of Adam Smith was not dissimilar in its relation to the Reign of Law. It was his labour to prove that in the rude contrivances of Legislation, due account had not been taken of the natural forces with which it had to deal. He showed that among the very elements of human character there were instincts, and desires, and faculties of contrivance, all of which by clumsy machinery had been impeded, and obstructed, and diverted from the channels in which they ought to work. He could not, however, test his reasoning as the Inventor could, by continual experiment. He had to rely on abstract reasoning, and on such verification as could be drawn from the complicated phenomena of the Body Politic. In this respect the work of Adam Smith was harder than the work of Watt. And why it was harder is a question which it may be well to ask. It is not surprising that the methods of applying to our own use the Powers of external Nature, should be matter of difficult research. But it may well seem strange that the forces which have their seat within ourselves—in the Mind and Character of Man—should be so unknown to us as to require careful reasoning and observation before we know how to use them with success for the attainment of our ends. Yet so it is. The conscious energies of the Will are ever tempted to march directly upon objects which can only be reached by circuitous methods of approach. And so the Wealth of Nations, and the skill of Crafts, and the success of Trade, had all been hindered by the measures designed for their protection. The promptings of individual interest had been checked and thwarted and driven into channels less fruitful than those which they would have naturally found.

On the other hand, the discovery of the Steam Engine, like every other weapon placed at the disposal of Mind, gave a new stimulus to the motives, and a new form to the conditions, by which the conduct of thousands was determined. Little did the brilliant Professor know that the discoveries of his humble friend would yet, in their results, serve to limit the conclusions of his own Philosophy. In the meantime, all that he knew of Watt and of his personal history seemed to be, and really was, a signal illustration of the follies of restriction. For no other reason than that he had not been born in Glasgow, Watt could not legally sell the products of his ingenuity and labour in that City. The spirit and the laws of corporate monopoly rigidly excluded him; and the company of "Hammermen" insisted on the exclusion being maintained, for fear of "loss and skaith to the Burgesses and Craftsmen of Glasgow, by the intrusion of strangers."[1] The working-classes themselves were among the most strenuous supporters of a system which diminished the value by restricting the area of their labour. Fortunately the University had privileges of its own, which, within its own property, excluded the jurisdiction of a Municipality and a Craft not more ignorant or more selfish than their contemporaries at the time. It may well be supposed, that Adam Smith's opinions on freedom of labour must have been influenced by personal observation of the working of such laws in the case of a man who, though still obscure, was even then appreciated by those who knew him for ingenuity and resource.

1 Smiles' "Life of Watt," p. 105.

In looking at restrictions such as these, there was nothing then to suggest to Adam Smith the consequences which might arise from the entire freedom of labour, when that labour was placed under new conditions. He had no knowledge, and he could then have no conception, what these new conditions were to be. Yet they were being silently prepared and determined in the very years in which he spoke and wrote. His friend Watt was a principal agent in the great impending change. But Watt was not alone. Other minds were working at the same time whose labours were to match with a curious fittingness into his. Indeed, the work which was going on in those years is only one example of a law of which many other examples may be found. It is an order of facts observable in the progress of Mankind, that long ages of comparative silence and inaction are broken up, and brought to an end, by shorter periods of almost preternatural activity. And that activity is generally spent in paths of investigation, which, though independent, are converging. Different minds, pursuing different lines of thought, find themselves meeting upon common ground. Such, in respect to literature, was the period of the Revival of Learning: such, in respect to Religion, was the period of the Reformation: such, in respect to the abstract sciences, was the period of Tycho Brahe, of Galileo, and of Kepler. Hardly less memorable than these, certainly not less powerful, as affecting the condition of society, were those few years in the last quarter of the eighteenth century, which were marked by such an extraordinary burst of Mechanical Invention. Hargreaves, and Arkwright, and Watt, and Crompton, and Cartwright, were all contemporaries. They were all working at the same time, and in the same direction. Out of their inventions there arose for the first time what is now known as the Factory system; and out of the Factory system arose a condition of things as affecting human labour, which was entirely new in the history of the world. The change thus effected is a signal illustration of the relation in which Natural Law stands to Positive Institution in the realm of Mind. Let us look for a moment at its history and results.

The Common Law of England had placed no restrictions upon labour. The only restrictions which existed arose either from the special monopolies of Corporate Bodies, or from the General Statute of Apprenticeship. This statute had been passed in the reign of Elizabeth. It provided that no man should work at any craft on his own account until he had served an apprenticeship of seven years. But the Statute of Apprenticeship being in derogation of common rights, had always been construed strictly by the Courts of Law; and so it had come to pass that two great rules of limitation had been applied to it. First, it was held to apply only to such crafts of skill as were known at the time of its being passed; and secondly, it was held not to apply at all to rural districts, but only to market towns. From these two rules of limitation, it resulted, first, that all trades and employments were free which had arisen since the commencement of the seventeenth century, and, secondly, that even the older crafts were free also if they were prosecuted outside the boundaries of towns.

Such was the condition of the law when the inventions of Adam Smith's contemporaries brought into existence employments which were entirely new, and opened them to that unrestricted competition, the advantage of which he had laid down as a universal doctrine.

Spinning and weaving were not new. They were as old as the memory of Mankind. But the simple mechanism by which these arts were prosecuted were almost equally old, and had undergone little change and little improvement. In 1760 the Spinning-Wheel, and the common Loom, as used by the people of Yorkshire, were little in advance of the implements for the same purpose which had been in use beyond the reach of History. The Spindle which is depicted on the monuments of Egypt was, until a few years ago, familiar in the Highlands. The essential feature of this ancient industry, so far as its effects upon social conditions are concerned, was that it was separate and not gregarious. It did not interfere with, but rather was congenial to, Family Life, for thousands of years,

> "Maids at the Wheel, the Weaver at his Loom,
> Sat blithe and happy."[1]

But the pressure of new necessities had arisen, and these could be met only by new inventions. Towards the middle of the eighteenth century, the greatest difficulty was experienced by weavers and spinners in England in maintaining their position in the markets of the world. It is curious how each new mechanical invention gave rise to the necessities out of which the next arose. The invention of the Fly Shuttle in weaving, so early as 1733, seems to have given the first impulse to all that followed. By means of this invention the power of weaving overtook the power of spinning. An adequate supply of yarn could not be procured under the ancient methods of that most ancient industry. New conditions gave rise to new motives, and new motives called into play the latent energies of Mind. The time and the cost of collecting the products of so many scattered labourers enhanced unduly the cost of manufacture, and

[1] Wordsworth's noble sonnet—

"Nuns fret not at their convent's narrow room."

even when the remuneration was reduced to the lowest point compatible with existence, that cost was still too high. Something was imperatively required to economise the work of human hands—some more elaborate contrivance to make that work go further than before. And so Hargreaves' invention arose, not before the time.[1] And when his Spinning Jenny had been invented, a still more elaborate and powerful combination of mechanical adjustments was soon perfected in the hands of Arkwright.[2] When the Spinning Frame was invented, and when Crompton's farther invention of the Mule Jenny speedily followed,[3] the new order of things had been fairly inaugurated. The great change had come, and the survivance of the ancient domestic industries of so many centuries was no longer possible.

And just as Hargreaves and Arkwright and Crompton were inventing the new machines which were to be moved, Watt was labouring at the new Power which was to move them. But meanwhile before the Steam Engine had been made available, the Factory system had begun under the old motive-power of Water; and here it is very curious to observe how each stage in the progress of discovery had, by way of natural consequence, its own special effect on the conduct and the Wills of men. Very soon the course of every moun-

[1] 1765-7. [2] 1769-71. [3] 1787.

tain stream in Lancashire and Yorkshire, was marked by Factories. This again had another consequence. It was a necessity of the case that such Factories must generally be situated at a distance from pre-existing populations, and, therefore, from a full supply of labour. Consequently they had to create communities for themselves. From this necessity, again, it arose that the earlier mills were worked under a system of Apprenticeship. The due attendance of the requisite number of "hands" was secured by engagements which bound the labourer to his work for a definite period.

And now for the first time appeared some of the consequences of gregarious labour under the working of Natural Laws, and under no restrictions from Positive Institution. The millowners collected as Apprentices boys and girls, and youths and men, and women, of all ages. In very many cases no provision adequate, or even decent, was provided for their accommodation. The hours of labour were excessive. The ceaseless and untiring agency of machines kept no reckoning of the exhaustion of human nerves. The Factory system had not been many years in operation when its effects were seen. A whole generation were growing up under conditions of physical degeneracy, of mental ignorance, and of moral corruption.

The first public man to bring it under the notice of

Parliament with a view to remedy, was, to his immortal honour, a master manufacturer, to whom the new industry had brought wealth, and power, and station. In 1802 the elder Sir Robert Peel was the first to introduce a bill to interfere by law with the natural effects of unrestricted competition in human labour. It is characteristic of the slow progress of new ideas in the English mind, and of its strong instinct to adopt no measure which does not stand in some clear relation to pre-existing laws, that Sir Robert Peel's bill was limited strictly to the regulation of the labour of Apprentices. Children and young persons who were not Apprentices might be subject to the same evils, but for them no remedy was asked or provided. The notion was, that as Apprentices were already under Statutory provisions, and were subjects of a legal contract, it was permissible that their hours of labour should be regulated by positive enactment. But the Parliament which was familiar with restrictions on the products of labour, and with restrictions of monopoly on labour itself—which restrictions were for the purpose of securing supposed economic benefits, would not listen to any proposal to regulate "free" labour for the purpose of avoiding even the most frightful moral evils. These evils, however great they might be, were the result of "natural laws," and were incident to the personal freedom of Employers and

Employed. In the case of Apprentices, however, it was conceded that restriction might be tolerated. And so through this narrow door the first of the Factory Acts was passed. It is a history which illustrates, in the clearest light, the sense in which human conduct, both individually and collectively, is determined by Natural Law. If Watt's Steam Engine had been invented earlier—if mills had not been at first erected away from the centres of population, in order to follow the course of streams—if consequently the evils of the Factory system had not begun to be observable in the labour of Apprentices, there is no saying how much longer those evils might have been allowed to fester without even an assertion of the right to check them. The Act of 1802,[1] though useless in every other sense, was invaluable at least in making this assertion.

Meanwhile Watt's great invention had been completed. And now a new cycle of events began, arising by way of natural consequence out of the Reign of Law. When the perfected Steam Engine became applicable to mills, it was no longer always cheaper to erect them in rural districts; on the contrary, it was often cheaper to have them in the towns, near a full supply of labour, and a cheap supply of fuel. With this change came the abandon-

[1] 42 and 43 Geo. III., cap. 73.

ment of the system of Apprenticeship. It was now "free" labour which more and more supplied the mills. But this only led to the same evils in an aggravated form. Children and women were especially valuable in the work of mills. There were parts of the machinery which might be fed by almost infant "hands." The earnings of children became an irresistible temptation to the parents. They were sent to the factory at the earliest age, and they worked during the whole hours that the machinery was kept at work. The result of this system was soon apparent. In 1815, thirteen years after he had obtained the Act of 1802, Sir Robert Peel came back to Parliament and told them that the former Act had become useless—that mills were now generally worked, not by water, but by steam—that Apprentices had been given up, but that the same exhausting and demoralising labour, from which Parliament had intended to relieve Apprentices, was the lot of thousands and thousands of the children of the free poor. In the following year, 1816, pressing upon the House of Commons a new measure of restriction, he added, that unless the Legislature extended to these children the same protection which it had intended to afford to the Apprentice class, it had come to this—that the great mechanical inventions which were the glory of the age would be a curse rather than a blessing to the country.

These were strong words from a master manufacturer; but they were not more strong than true.[1]

Thus began that great debate which in principle may be said to be not ended yet:—the debate, how far it is legitimate or wise in Positive Institution to interfere for moral ends with the freedom of the individual Will? Cobbett denounced the opposition to restrictive measures as a contest of "Mammon against Mercy." No doubt personal interests were strong in the forming of opinion, and some indignation was natural against those who seemed to regard the absolute neglect of a whole generation, and the total abandonment of them to the debasing effects of excessive toil, as nothing compared with the slightest check on the accumulations of the Warehouse. But the opposition was not in the main due either to selfishness or indifference. False intellectual conceptions—false views both of principle and of fact—were its real foundation. Some of the ablest men in Parliament, who were wholly unaffected by any bias of personal interest, declared that nothing would induce them to interfere with the labour which they called "free." Had not the working classes a right to employ their children as they pleased? Who were better able to judge than fathers and mothers of the capacities

[1] "Hansard Parl. Deb." vols. xxxi. and xxxiii.—Sir Robert's Speech on Motion for a Committee, April 3, 1816.

of their children? Why interfere for the protection of those who already had the best and most natural of all protections? Such were some of the arguments against interfering with free labour.

Now in what sense was this labour free? It was free from legal compulsion—that is to say, it was free from that kind of compulsion which arises out of the public Will of the whole community imposed by authority upon the conduct of individuals. But there was another kind of force from which this labour was not free—the force of overpowering motive operating on the Will of the labourers themselves. If one parent, more careful than others of the welfare of his children, and moved less exclusively by the desire of gain, withdrew his children at an earlier hour than others from Factory Work, his children were liable to be dismissed and not employed at all.[1] On the other hand, motives hardly less powerful were in constant operation on the masters. The ceaseless, and increasing, and unrestricted competition amongst themselves,—the eagerness with which human energies rush into new openings for capital, for enterprise, and for skill,—made them, as a class, insensible to the frightful evils which were arising from that com-

[1] This was very forcibly explained, both by Sir Robert and by his son, Mr. Peel, in the debate of Feb. 23, 1818.

petition for the means of subsistence which is the impelling motive of labour.

Nor were there wanting arguments, founded on the constancy of Natural Laws, against any attempt on the part of Legislative authority to interfere with the "freedom" of individual Will. The competition between the possessors of capital was a competition not confined to our own country. It was also an international competition. In Belgium especially, and in other countries, there was the same rush along the new paths of industry. If the children's hours of labour were curtailed, it would involve of necessity a curtailment also of the adult labour, which would not be available when left alone. This would be a curtailment of the working time of the whole mill; and this would involve a corresponding reduction of the produce. No similar reduction of produce would arise in Foreign mills. In competition with them the margin of profit was already small. The diminution of produce from restricted labour would destroy that margin. Capital would be driven to countries where labour was still free from such restrictions, and the result would be more fatal to the interest of the working classes of the English towns than any of the results arising from the existing hours of work. All these consequences were represented as inevitable. They must arise out of the operation of invariable laws,

Such were the arguments—urged in every variety of form, and supported by every kind of statistical detail —by which the first Factory Acts were vehemently opposed.

And, indeed, in looking back at the debates of that time, we cannot fail to see that the reasoning of those who opposed restriction on Free Labour met with no adequate reply. Not only were the supporters of restriction hampered by a desire to keep their conclusions within the scope of a very limited measure; not only were they anxious to repudiate consequences which did legitimately follow from their own premises; but they were themselves really ignorant of the fundamental principles which were at issue in the strife. Their conclusions were arrived at through instincts of the heart. The pale faces of little children, stunted and outworn, carried them to their result across every difficulty of argument, and in defiance of the alleged opposition of inevitable laws. And yet, if the supporters of the Factory Acts had only known it, all true abstract argument on the subject was their own. The conclusions to which they pointed were as true in the light of Reason, as they felt them to be true in the light of Conscience.

The truth is, that some of the finest distinctions in Philosophy were then for the first time emerging on the stage of Politics. The newest debates of Parliament were circling unconsciously round one of the oldest disputations of the Schools. A question of practical legislation had arisen which involved one of the most difficult problems in metaphysical analysis. On the one hand, Freedom was asserted for the Will under conditions and in a sense in which it did not exist. On the other hand, Freedom was denied to the Will in a sense in which the instincts of humanity testified to its presence, and to the possibility of its being exerted with effect. The true Doctrine of Necessity was exemplified in the conduct of Employers and Employed—that conduct being determined in a wrong direction by the force of overpowering motives. The false Doctrine of Necessity was exemplified in the argument, that this conduct could not be changed under the force of higher motives asserting themselves through the Will of the Community in the form of Law.

The antagonism which was and still is so often assumed between Natural Law and Human Law, or in other words between Natural Law and Positive Institution, is an antagonism which may indeed exist, and does very often exist. But it is also an antagonism which may be eliminated, and must be eliminated, if Legislation is ever to be attended with permanent success. It is, alas, a Natural Law that men should be thoughtless, and selfish, and reckless of moral consequences, when they are bent exclusively on material results. But when the consequences of this conduct have been brought home to their convictions by the force of imminent danger or of actual calamity, it is a law not less natural that they should take alarm, that they should retrace their steps, and that by walking in another course they should bring about conditions of a better kind. The Laws of Man are also Laws of Nature, when founded on a true perception of natural tendencies and a just appreciation of combined results. On the other hand, Human Laws are at variance with, or antagonistic to the Laws of Nature, when founded either on the desire of attaining a wrong end, or on the attempt to reach a right end by mistaken means. In either of these cases Positive Institution and Natural Law become opposed, and thus a bad contrivance in Legislation, like a bad contrivance in mechanics, comes always to some dead-lock at last. Time and Natural Consequence are great Teachers in Politics as in other things. Our sins and our ignorances find us out. Both in conduct and in opinion Natural Law is ever working to convict error, to reveal and to confirm the truth.[1]

And so it was that the sad phenomena of Factory labour were beginning to indicate the great difference

1 "Opinionum enim commenta delet dies; naturæ judicia confirmat."—Cicero, "De Nat. Deor." lib. ii. c. 3.

between the results of perfect freedom of exchange in the products of labour and the results of perfect freedom of competition in Labour itself. Perhaps that difference ought to have been foreseen, for the cause of it is plain enough. There are certain results for the attainment of which the natural instincts of individual men not only may be trusted, but must be trusted as the best and indeed the only guide. There are other results of which as a rule those instincts will take no heed whatever, and for the attainment of which, if they are to be attained at all, the higher faculties of our nature must impose their Will in authoritative expressions of Human Law. In all that wide circle of operations which have for their immediate result the getting of wealth, there is a sagacity and a cunning in the instincts of labour and in the love of gain compared with which all legislative wisdom is ignorance and folly. But the instincts of labour, having for their conscious purpose the acquisition of wealth, are instincts which, under the stimulus and necessities of modern society, are blind to all other results whatever. They override even the love of life; they silence even the fear of death. Trades in which the labourers never reach beyond middle life—trades in which the work is uniformly fatal within a few years—trades in which those who follow them are liable to loathsome and torturing disease—all are filled by the enlistment of an unfailing series of recruits. If, therefore, there be some things desirable or needful for a Community other than the acquisition of wealth,—if mental ignorance, and physical degeneracy, be evils dangerous to social and political prosperity, then these results cannot and must not be trusted to the instincts of individual men. And why? Because the few motives which bear upon them, and which consequently determine their conduct, have become almost as imperious as the motives which determine the conduct of the lower animals. Observers whose duties have called them to a close investigation of the facts, have never failed to be impressed with those facts as the result of Laws against which the individual Will is unable to contend. Overpowering motives arise out of the conditions of society—out of the force of habit—out of the helplessness of poverty—out of the thoughtlessness of wealth—out of the eagerness of competition—out of the very virtues even of industrial skill. These constitute an aggregate of power tending in one direction, which make the resulting action of Mind as certain as the action of Inanimate Force. "Thus," says Mr. Baker, one of the most experienced of our Factory Inspectors, "most of the workshops of this great commercial country are found to have fallen into *the inevitable track of competitive industry*, when unrestricted by law,—namely, to cheapen prices by the employment of women and children in the first instance, and then to increase production by protracted hours of work, without much regard to age, to sex, or to physical capability." This is the result of Nature—of Nature, at least, such as ours now is. But it is the result of that Nature with all its nobler powers allowed to sleep. Power to control such evils has been given to Man, and he is bound to use it. "Free labour, even in a free country," as Mr. Baker says, "requires the strong arm of the law to protect it from the cupidity and ignorance of parents."[1] And by the "strong arm of the law" is meant nothing but the law of Conscience and of Reason asserting itself over the lower instincts of our nature. If under such conditions of society, higher motives are ever to prevail, they must be supplied from without, and must be imposed in authoritative form through the legitimate organs of Positive Institution.[2]

And so the Factory Acts instead of being excused as exceptional, and pleaded for as justified only under extraordinary conditions, ought to be recognised as in truth the first Legislative recognition of a great Natural Law,

1 "Reports of the Inspectors of Factories, half-year Oct. 1864," p. 84.

2 Bad as the consequences were of individual freedom under unrestricted competition in the case of labour in factories, the results were still more horrible in the case of labour in mines. In 1842 it was found absolutely necessary to prohibit altogether the labour of women and young children in mines and collieries.

quite as important as Freedom of Trade, and which like this last, was yet destined to claim for itself wider and wider application.

Accordingly, since the year when the first Sir Robert Peel pleaded the cause of Factory Apprentices, there has been going on a double movement in Legislation, one a movement of retreat, the other a movement of advance. Step by step Legislation has retired from a Province once considered peculiarly its own: step by step it has advanced into another Province within which the Schools of Political Economy would have denied it a foot of ground. Since 1802, there have been passed a long series of laws removing, one after another, all restrictions which aimed at guiding the individual Will in its sharp and sagacious pursuit of material wealth. During the same period there have been passed another long series of Acts imposing restrictions more and more stringent on the individual Will in its blind and reckless disregard of moral ends.[1] In neither of these movements was Par-

[1] It was not till 1819 that Sir Robert Peel succeeded in passing an Act restricting the labour of unapprenticed children. This Act (59 Geo. III. c. 66) is therefore, properly speaking, the first of the Factory Acts—the first which affirmed the principle of restriction as legitimately applicable to "Free" Labour. But this, as well as a subsequent Act passed in 1825, at the instance of Sir J. Hobhouse, were practically inoperative from defective enforcing clauses. It was thus apparent that the State must charge itself not only with

liament impelled by the light of reason, but under the blessed teaching which belongs to the Reign of Law. False theory and mistaken conduct have been found out by the working of Natural Consequence. The abstract reasonings of Adam Smith had indeed long before prepared the minds of a few to perceive the true theory of unrestricted competition in the interchange of goods. But as it needed the practical results of restriction—distress, discontent, and the danger of civil commotion—to bring home to the national understanding the economic error of the old commercial systems; so also as regards the grievous results of unrestricted competition in human labour, our only effective teaching has been that of hard experience. The doctrines of Adam Smith, when applied here, were a hindrance and not a help. The Political Economists were, almost to a man, hostile to restrictive legislation. They did not see what would be the working of Natural Law upon the Human Will, when that Will was exposed to overpowering motives under debased conditions of understanding and of heart. They did not

laying down the law, but also with the duty of seeing it obeyed.' It was not till this great question was taken in hand by Lord Ashley that any effectual measure was passed. His Bill became Law in 1833, as 3 and 4 Will. IV. c. 103. Nothing but a stringent system of Government Inspection was of any avail against the powerful combination of motives, out of which the evils of the Factory system arose.

see the higher Law which Parliament was asserting when it was driven by sheer instinctive horror of actual results, to prohibit "free" labourers from disposing as they pleased of the labour of their children.

To this hour the principle on which this great counter-movement rests as regards our ideas of the legitimate province of Legislation, has never been philosophically treated. The Laws on which it depends, and which it does but recognise, have never been scientifically defined. We are still in a state of tutelage—advancing with slow and reluctant steps[1] in the path indicated by the teachings of Natural Consequence. The last Report on the

[1] The steps here referred to are certainly becoming every year less slow and less reluctant. Since this work was published, the "Factory Acts' Extension Act" of 1867 has extended the provisions of those Acts to all establishments which employ fifty persons; and the "Workshop Regulation Act" of the same year, has carried the protection of the law into the precincts of "any room or place whatever in which any handicraft is carried on." Nay more, it extends that protection even to children who are working, not for wages at all, but only "under a parent." The principle of "State interference" is here carried to its utmost length. It is characteristic of the cautious and tentative character of English Legislation that it becomes gradually committed to great general principles, not through any perception of the truth and value of those principles in the abstract, but gradually, and through the compulsion of particular necessities. And to the last possible moment the general application of such principles is always resisted. But no argument can be used in favour of compulsory education, as regards children in "workshops," which is not equally applicable to all children whatever.

Employment of Children shows that evils as bad as ever existed before the passing of the Factory Acts, prevail at this moment among large classes of our operative population, and demand again, as imperatively as before, an authoritative interference of Positive Institution with the freedom of the individual Will. The fact of such legislation has indeed gained a sort of silent acquiescence, and some of the old opponents have admitted that their fear of the results, in an economical point of view, has proved erroneous. But there is still no clear and well-grounded intellectual perception of the deep foundations of principle on which it rests. Nor is there among a large section of Politicians any adequate appreciation of the powerful influence it has had in improving the physical condition of the people, and securing their contentment with the Laws under which they live.

When, however, we think for a moment of the frightful nature of the evils which this Legislation has checked, and which to a large extent it has remedied—when we recollect the inevitable connexion between suffering and political disaffection—when we consider the great moral laws which were being trodden under foot from mere thoughtlessness and greediness—we shall be convinced that if, during the last fifty years, it has been given to this country to make any progress in Political Science, that progress has been in nothing happier than in the Factory

Legislation. The names of those who strove for it, and through whose faith and perseverance it was ultimately carried, are, and ever will be, in the history of Politics, immortal names. No Government and no Minister has ever done a greater—perhaps, all things considered, none has ever done so great a service. It was altogether a new era in Legislation—the adoption of a new principle—the establishment of a new idea. Nor is that principle and that idea even now thoroughly understood. The promptings of individual self-interest are still relied upon for the accomplishment of good which it does not belong to them even to suggest, and which they can never be trusted to pursue. Proposals for legislative interference with a view to arrest some of the most frightful evils of Society, are still constantly opposed, not by careful analysis of their tendency, but by general assertions of Natural Law as opposed to all legislation of the kind. "You cannot make men moral by Act of Parliament"—such is a common enunciation of Principle, which, like many others of the same kind, is in one sense a truism, and in every other sense a fallacy. It is true that neither wealth, nor health, nor knowledge, nor morality, can be given by Act of Parliament. But it is also true that the acquisition of one and of all of these can be impeded and prevented by bad laws, as well as aided and encouraged by wise and appropriate legislation.

There is no doctrine in Physics more certainly true than this doctrine in Politics—that every practice which the authority of Society recognises or supports has its own train of consequences, which, for evil or for good, can be modified or changed in an infinite variety of degrees according as that sanction is given or withheld. Innumerable illustrations of this truth will arise wherever we take the trouble to trace any social or political phenomena through the sequences of cause and effect from which they come. Not unfrequently these illustrations are of a melancholy kind, and give us much to think of respecting the better understanding and the better management of our complicated nature. Thus, for example, there seems good reason to believe there is a direct relation between the amount of life and property annually sacrificed by shipwreck, and the legislation which recognises and sanctions Insurance to the full amount of the value of ship and cargo. The cause of this is obvious. Care for life is less eager and less wakeful than care for property. This is true even when men are dealing equally with their own property, and with their own lives. It is still more true when they are dealing not only with property which is their own, but with lives which belong to others. The inevitable effect of such Insurance is therefore to relax the motives of self-interest, which are the strongest incitements to precaution.[1] Similar results appear in a thousand other cases, both of laws still existing, and of laws which have been repealed. The conduct of men depends on the balance of motives which are brought to bear upon them. In supplying those motives, external conditions and mental character act and react upon each other. Both of these can be affected, and affected powerfully, by Positive Institution.

The restraints of Positive Institution are not, however, the only means,—very often they are not the best means by which to lighten the overpowering pressure of particular motives upon the individual Will. For as the Reason and the Conscience of the whole Political Community can interfere by the exercise of authority, so also may adequate remedies be found in the reason and the conscience of Voluntary Societies. The external conditions which tell upon the individual Will are themselves very often nothing but conditions depending on the aggregate Will of those around us; and if upon them, by any means, new motives can be brought to bear, then the whole of those external conditions may be changed. The language which is used in the name of Economic Science constantly involves in this matter

[1] A curious and instructive Paper upon this subject has been published by Mr. Edwin Chadwick, having been read before a recent meeting of the Social Science Association.

the same fallacy which has already been pointed out in the language used in the name of Physical Science. It is often said that the conduct and condition of men are governed by invariable laws; and the conclusion is that the evils which arise by way of natural consequence out of the action of those laws, are evils against which the struggles of the Will are hopeless. But the facts on which this conclusion is founded, are, as usual, inaccurately stated. The conditions of human life and conduct, like the conditions of all natural phenomena, are never governed by those separate and individual forces which alone are invariable, but always by combinations among those forces—which combinations are of endless variety, and of endless capability of change. Different motives arise out of the inborn gifts of character, and out of the conditions of external circumstance. It is true, indeed, that there are in the mind of Man, as there are in Nature, certain forces originally implanted, which are unchangeable in this sense, that they have an invariable tendency to determine conduct in a particular direction. But as in Nature we have a power of commanding her elementary forces by the methods of adjustment, so in the Realm of Mind we can operate on the same principle, by setting one motive to counteract another: and by combination among many motives, we can influence in a degree, and to an extent as yet unknown, the conduct and the condition of Mankind.

Nor are the resources of Contrivance limited to adjustment among the motives which arise only out of existing conditions. New motives can be evoked and put in action by the adopting of appropriate means. The mere founding, for example, of a Voluntary Society for any given purpose, evolves out of the primary elements of human character a latent force of the most powerful kind; namely, the motive—the sentiment—the feeling—the passion as it often is, of the Spirit of Association. This is a passion which defies analysis. The cynic may reduce it to a form of selfishness—and undoubtedly the identification of the interests and the desires of Self with the Society for which this passion is conceived, lies at its very root and is of its very essence. It is true, also, that it is a passion so powerful as to need strong control—without which control it generates some of the very meanest emotions of the heart. Out of it there has come, and there comes again and again from age to age, a spirit of hatred even against good itself, when that good is the work of any one who "followeth not us." It is a force, nevertheless, rooted in the Nature of Man, implanted there as part of its constitution, and like all others of this character, given him for a purpose, and having its own legitimate field of operation. Nor is that field a narrow one. The spirit of Association is the fountain of much that is noblest in human character, and of much that is most heroic in human conduct. For all the desires and aspirations of Self are not selfish. The interests of Self, justly appreciated and rightly understood, may be, nay indeed must be, the interests also of other men—of Society—of Country—of the Church, and of the World.

And so it is that when the aim of any given Association is a high aim, directed to ends really good, and seeking the attainment of them by just methods of procedure, the spirit it evokes becomes itself a new "Law" —a special force operating powerfully for good on the mind of every individual subject to its influence. Some pre-existing motives it modifies—some it neutralises—some it suppresses altogether—some it compels to work in new directions. But in all cases the Spirit of Association is in itself a power—a force—a Law in the Realm of Mind. What it can do, and what it cannot do, in affecting the conditions of Society, is a problem not to be solved so easily and so summarily as some dogmatists in political philosophy would have us to believe. It is a question which, like so many others, is not likely to be solved by abstract reasoning without the help of actual experiment. And this experiment is being tried. The instincts of men, truer often than the conclusions of philosophy, have rebelled against the doctrine that they are the sport of circumstances. Yet finding by hard experience that this is often true of the individual Will when standing by itself, they have resolved to try whether it is equally true of the Collective Will, guided by the spirit and strengthened under the discipline of Association. Hence the phenomena of Combination as a means of affecting the condition of labour—phenomena so alarming to many minds, and certainly so well deserving of attention. Let us look for a moment at the important illustrations of the Reign of Law which these phenomena afford.

A moment's consideration will convince us that the same necessities of labour which were found to determine so fatally the condition of women and children, are necessities which apply without any abatement to the labour of adult men. They must be subject to the same pressure of inducements. Nay, more, it is only through them that this pressure can reach the women who are their wives, and the children who are their children. If overpowering motives did not equally determine the conduct and condition of adult men, no legislation would have been required for the protection of their families. If a man is placed under such conditions that he cannot save his wife and child from exhausting labour, it is certain that the same conditions will impose a like necessity upon himself. Nevertheless, Parliament has resolutely and wisely refused to interfere on his behalf. And why? Because the argument is that the adult man is able, or ought to be able, to defend himself. And so he can; but how? Only by combination. The "law" which results in excessive labour is the law of competition—that is, it is the attraction exerted upon the Wills of a multitude of individual men by the rewards of labour. The pressure of this attraction can only be lightened by bringing those Wills under the power of counter motives, which may induce them to postpone, to some higher interest, the immediate appetites of gain. And this is the work which Combination does. It comes in the place of Positive Institution. Those who are under it "are a Law unto themselves."

Nor is it unimportant to observe that what Combination does for the protection of labour, it does better, and with better consequences, than Positive Institution can ever do. Men are driven to excessive labour, because if they don't work excessively, others will. But it is the effect of Combination that others won't. Under Positive Institution they are not allowed. Under Combination they are determined not. And as the forming of an intelligent resolution, and the abiding by it, are higher exercises of mind than the mere passive obedience to authority, so is the good effected by Combination a higher good than that resulting from Factory Legislation. It tends to form and to strengthen character. It tends to subordinate the present to the future, and the temporary interests of Self to the permanent welfare of a Brotherhood of men. And this it tends to do in classes otherwise prone to follow only the impulse of the moment, and to consider only the apparent interests of the individual.

These considerations should disabuse our minds of the unjust and unreasonable prejudice against the principle of Combination which still betrays itself so strongly in the language of many politicians. When the Working Classes combine for the protection of their own labour against the effects of unrestricted competition, they are simply taking that course which is recommended alike by reason and by experience. It is the course which Parliament has indicated as the right course both by what it has itself done, and by what it has declined to do. Nor can there be any greater mistake than to suppose that this course involves necessarily any rebellion against the laws of economic science. Combination is an appeal to the most fundamental of all Natural Laws—to the law of Contrivance—to the power of Adjustment—wielding, through Reason and Conscience, the elementary forces of Human Character. Of the constancy and "invariability" of these no doubt or denial is involved. Rather the reverse. It is upon instinctive trust in that constancy that all social and political Contrivance rests. And so we need not be surprised to find that, through the organised efforts of communities of men, the evils which arise by way of natural consequence out of the helplessness and thoughtlessness of the individual Will, are evils which to a large extent can be met and overcome.

But though all this is true, universally, of the principle of Combination, it is very far from being true, universally, of the particular purposes to which Combination is applied. All the sources of error which have so long perverted Legislation, are equally powerful in perverting the aims, and in misdirecting the efforts of Voluntary Association. If the upper classes, with all the advantages of leisure, and of culture, and of learning, have been so unable, as we have seen them to be, to measure the effect of the laws they made, how much more must we expect errors and misconceptions of the most grievous kind to beset the action of those who—through poverty and ignorance, and often through much suffering—have been able to do little more than strike blindly against evils whose pressure they could feel, but whose root and remedy they could neither see nor understand?

Accordingly, the history of Combination among the Working Classes has, until a very recent period, been a sad history of misdirected effort—of strength put forth only in violence and disorder, and of the virtues of Brotherhood lost in tyrannical suppression of all individual freedom. Its heaviest blows have been often aimed at the most powerful agencies for good. One of the very earliest forms of Combination has been that which was directed against the introduction and improvements of machinery. The Working Classes have always encountered with jealousy and fear those triumphs of Mechanical Invention whose function it is to economise labour and to multiply the fruits of industry. It would be hard to blame them. What class is there which can say with truth that they have themselves been able always to follow with intelligent foresight the links of Natural Consequence through the darkness into which they so often lead? For almost every great step in the advance of civilization plunges at first through some passage which seems dangerous, or at least obscure. The happiest achievements of Contrivance have their own aspects of apparent danger, and their own real incidents of temporary evil. Every new machine displaces and disorganises pre-existing forms of labour; and we have seen that, even in its ultimate effects, the advance of Mechanical Invention developed new dangers to the Working Classes—dangers only to be avoided by measures which were not taken, and by precautions which were not adopted.

It would be well if, from the past convicted errors,

both of Legislation and of Combination, we could extract some conclusions of general principle capable of helping us in the difficulties of our own time. In looking at the root of those errors, it would seem that, in order to avoid them, two things are necessary—first, unshaken faith in great Natural Laws; and, secondly, a faith not less assured in the free agency of Man to secure by appropriate means the working of those laws for good. Thus, the love of gain is an instinct implanted in the human mind, and the endeavour to suppress it has always been the violation of a Natural Law. In like manner, Mechanical Invention is a Law of Nature in the highest and strictest sense. The power of it and the love of it are among the elementary forces of human character. Each fresh exertion of it is, and must be, according to the constitution and course of Nature—leading to higher and higher fulfilments of the original Purpose of Man's Creation, which was, that he should not only inhabit the Earth, as the beasts inhabit it, but that he should subdue it.

So also combination is natural to Man. The desire for it and the need of it, grow with the growth of knowledge and with the increasing complications of Society. It has now, for the most part, emerged from the stage of rude ignorance which led to the breaking of machinery. It is conducted, comparatively at least, with high intelligence, and aims for the most part at legitimate objects of desire. Yet in the rebellion which has been roused against the doctrines of Necessity, founded on false conceptions of Invariable Law, there is a constant danger lest the Spirit of Association should attempt to act against Nature instead of acting with it. There is, for example, a Law—an observed order of facts—in respect to Man, which the Working Classes too often forget, but which can neither be violated nor neglected with impunity. That Law is the Law of inequality—the various degrees in which the gifts both of Body and of Mind are shared among men. This is one of the most fundamental facts of human nature. Nor is it difficult to see how it should be also one of the most beneficent. But it is a fact against which the spirit of Combination is very apt to assume an attitude of permanent insurrection. It is, of course, the business and the function of Combination to subordinate in some things the Individual to the Class; and the temptation is to make that subordination exclusive and complete. Hence the jealousy so often shown of wages measured by the amount of work performed. This is a jealousy of the superiority of reward which is naturally due to superiority of power, of industry, or of skill. But these are things which God has joined together, and which no man or combination of men have a right to put asunder. It is a marriage made in the morning of the world, and in every step of human progress we see its blessing and its fruit. If it be stupid to break machines and to proscribe Mechanical Invention, it is not less stupid to be jealous of this primeval adjustment between the varying energies of human character and the varying results which they are competent to attain.

This is not the place to enter in detail on the difficult and complicated question as to the limits within which Combinations can, and beyond which they cannot, affect the rewards of labour. They have certainly succeeded in limiting the hours of labour in cases where Legislation could not well have interfered;[1] and wherever the hours of labour are reduced without a corresponding reduction in wages, a substantial economic advantage is unquestionably secured. Equal confidence is expressed by many Associations, that as a matter of experience and of fact, they have succeeded in establishing higher rates of

1 Of this the Baking Trade is a good example. The hours of adult labour in this trade, under the effects of unrestricted competition, had come to be most injurious and oppressive. In Glasgow and in Edinburgh this condition of things has been effectually remedied by a Combination, whose exertions were successful, without (I believe) resort being ever had to the extreme measure of a Strike. The Baking Trade in London is still afflicted by the same oppressive hours of labour, because of the difficulty which has hitherto been experienced in organising there any Combination equally complete.

wages than would have accrued under the system of unrestricted competition. This may very well be true. It is a truth which casts no doubt whatever on the invariability of Economic Laws when these are rightly understood. They are invariable in the same sense, and in no other, in which all other Natural Laws are invariable. That is to say, they represent tendencies in human character determined by motives, which tendencies are constant, and may surely be relied on as producing always, under like conditions, their own appropriate effects. It is upon this constancy that Combination must rely for any power it can ever have: and it is the same constancy in the action of specific motives which sets bounds to the power of Combination, beyond which it can never pass. The same motive which impels the Workman to secure an adequate reward for his labour, impels the Manufacturer or the Trader to secure an adequate reward for his capital, his knowledge, and his skill. And although the desire of gain is not the only motive, and is often not the strongest motive, which impels men to persevere in enterprises once begun, yet if Combinations of Workmen should attempt to raise wages so high as to trench upon the minimum rate of profit which will induce men to carry on any given trade, then by a natural consequence, not less certain than any other, capital and enterprise and skill will be withdrawn from that trade, and those who depend upon it will be the first to suffer. Short, however, of this extreme result, there is generally a margin of ground upon which Combination may act with more or less effect. It may prevent arbitrary or capricious changes; and as there are practically many impediments in the way of men moving their capital from one employment to another, Combination may compel them to submit to lower rates of profit than would otherwise content them if those difficulties did not exist.

But to all these possibilities of influence there is a limit in the nature of things—in Natural Laws—that is, in the new motives which are brought into operation by new conditions. What that limit is, it must always be difficult to determine except by actual experiment. It is enough here to observe that in this, as in every other department of human conduct, men are being led gradually to a knowledge of the truth by the teachings of Natural Consequence. It is by the experience of actual results that we are taught both as to the objects which are legitimate objects of desire, and as to the proper methods by which these may be attained. The very attempt of the Working Classes to govern through Combination their own affairs, and to determine their own condition, is an Education in itself. On the extended scale in which that attempt is being made, it must accustom them to consider great general causes, and to estimate the manner and the degree in which these can be effected by the methods of Adjustment. Last, not least, it must lead them to study and to recognise the moral duties which are indeed the most fundamental of all Natural Laws. For it ought to be remembered, that the first and most important object of Combinations is one against which there can be no opposition founded on the doctrines of Economic Science. That object is to secure for the Working Classes those provisions against misfortune, sickness, accident, and age, which are amongst the first duties of all organised societies of men. How far through such agency the causes of pauperism may be successfully attacked, is a question on which we are only entering. In like manner, the conditions and limitations under which Combination may succeed in blending the functions and in uniting the profits of Capital and of Labour—this also is a question to be determined by Natural Laws, not yet fully explored or understood. But enough is known, and results sufficiently determinate have already been secured, to convince us that in this great department of Natural Law, as in every other, the Will of Man is not powerless when its energies are directed by wisdom, and when the choice of its methods is founded upon knowledge.

This is, indeed, the great lesson to be learnt from every inquiry into the constitution and course of things. Nature is a great Armoury of weapons, and implements, for the service and the use of Will. Many of them are too ponderous for Man to wield. He can only look with awe on the tremendous Forces which are everywhere seen yoked under the conditions of Adjustment—on the smoothness of their motions,—on the magnitude and the minuteness, on the silence and the perfection, of their work. But there are also many weapons hung upon the walls which lend themselves to human hands —lesser tools which Man can use. He cannot alter or modify them in shape or pattern—in quality, or in power. The fashion of them and the nature of them are fixed for ever. These are, indeed, invariable. Only if we know how to use them, then that use is ours. Then also the lesser contrivances which we can set in motion are ever found to work in perfect harmony with the vaster mechanisms which are moving overhead. And as in the material world no effort gives so fully the sense of work achieved as the subjugation of some Natural Force under the command of Will, so in the world of Mind no triumphs of the Spirit are happier than those by which some natural tendency of Human Character is led to the accomplishment of a purpose which is wise and good. It is for the gaining of these triumphs that Man has been gifted with the desire of Knowledge, and with the sense of Right, and with the faculties of Contrivance. In such triumphs lie the aim and purpose of all Natural Laws—for these they were all established—for these they all work, whether by way of encouragement, or of restraint, or of retribution.

Nothing is more striking in the history of Discovery than the ages during which men have been blind to the suggestions of Natural Law—suggestions which now appear so obvious that we wonder how the interpretation of them could have been missed so long. It is very easy to feel this wonder concerning others; it is much more difficult to remember that the same wonder will certainly be felt concerning ourselves. Such as we now see to have been the position of former generations in respect to things which they failed to understand,—such, we may depend upon it, is precisely our own position in regard to innumerable phenomena now constantly passing before our eyes. We may be sure of this; and we ought to be as glad of it as we are sure. For the world is not so prosperous or so happy as that we should readily or willingly believe in the exhaustion of the means which are at our disposal for its better guidance. Especially in the great Science of Politics, which investigates the complicated forces whose action and reaction determine the condition of Organised

Societies of men, we are still standing, as it were, only at the break of day. Our command over the external elements of Nature is, beyond all comparison, in advance of our command over the resources of Human Character.

Special causes retard the progress of knowledge in this department of inquiry. Many problems so difficult and intricate that they never can be solved except by patient observation, patient thought, and yet more patient action, are as yet hardly recognised to be problems at all. We look on the facts of Nature and of human life through the dulled eyes of Custom and Traditional Opinion. And when some misery worse than others forces itself upon the acknowledgment of the world, men are slow to discover or admit their own power over the sources whence such miseries come to be. That which is needed to open our eyes to such questions, is not mere intellectual power. Rarer and finer qualities have this work to do. Among the characteristics of the individual men who have exerted the most powerful influence for good on the condition of Society, no quality has been more remarkable than a certain natural openness and simplicity of mind. Readiness to entertain, willingness to accept, and enthusiasm to pursue, a new idea, have always been among the most fruitful gifts of genius.

Is it vain to hope that the thoughtfulness and candour which have been the natural inheritance of a few, may yet be more common among all educated men? The whole constitution and course of things would receive an earlier fulfilment did we carry about with us an habitual belief in the inexhaustible treasures which it holds—in the power of the agencies which it offers to Knowledge and Contrivance. For then the results of Natural Consequence would be accepted for that which they teach, and not simply submitted to for that which they inflict. The disorders of Society would not so often be supinely regarded as the result of inevitable laws, but would be seen as the fruit always of some ignorance or of some rebellion; and so the exhilarating conviction would be ours, that those disorders are within the reach of remedy through larger Knowledge and a better Will.

We hear much now of the "blessed light of Science;"[1] and if the methods and conditions of Physical inquiry were applied in a really philosophical spirit to Spiritual Phenomena, the influence of Science for good would be more powerful than it is. Meanwhile it is well to remember that although readiness to accept a new idea is essential to Discovery, it is equally true that new dangers beset and surround all new aspects of the truth.

1 See a remarkable passage in the concluding pages of "Ecce Homo."

Paradoxical as it may sound to say so, this is a consequence of the splendour of Man's endowments, of his freedom from direction,—of the swiftness and the subtlety of his mental powers. On her own narrow path Instinct is a surer guide than Reason, and accordingly it is often the higher faculties of the mind which are the most misleading. The Speculative Faculty is impatient of waiting upon Knowledge, and is ever as busy and as ingenious in finding out new paths of error as in supplying new interpretations of the truth. Hence in Philosophy the most extravagant errors have been constantly associated with the happiest intuitions, and it has remained for the successors of great men in another generation to separate their discoveries from their delusions. Hence also in Politics the great movements of Society have seldom been accomplished without raising many false interpretations of the Past, and many extravagant anticipations of the Future.[1] It cannot, indeed, be said with truth that the calamities of Nations have generally arisen from too great play being given to novel or theoretical conclusions. Rather the reverse. They have arisen, for the most part, from too little attention being paid to the progress of opinion, and to the insensible development of new conditions.

1 "Nos pères en 1789 ont été condamnés à passer des perspectives du Paradis aux scènes de l'Enfer."—Guizot, "L'Eglise et la Société," p. 218.

The question has been often raised, whether there is any law of growth, of progress, and of decay prevailing over Nations, as over individual Organisms. Whatever the solution of this question may be, it is certain that some causes are no longer in existence which produced—not indeed the corruption, but—the final overthrow of the great historical Nations of Antiquity. The epoch of conquering Races destroying the Governments, and reconstructing the Populations of the World, is an epoch which has passed away. Whatever causes there may be now of political decline are causes never brought to such rough detection, and never ending in catastrophes so complete. Yet, in modern days a condition of stagnation and decline has been the actual condition of many Political Societies for long periods of time. It is a condition prepared always by ignorance or neglect of some moral or economic laws, and determined by long-continued perseverance in a corresponding course of conduct. Then the laws which have been neglected assert themselves, and the retributions they inflict are indeed tremendous. In the last generation, and in our own time, the Old and the New Worlds have each afforded memorable examples of the Reign of Law over the course of Political events. Institutions maintained against the natural progress of Society have "foundered amidst fanatic storms." Other institutions upheld and cherished against justice, and humanity, and conscience, have yielded only to the scourge of War.

It is in the wake of such convulsions that reactions of opinion so often sweep over the Human Mind, as hurricanes sweep over the surface of the Sea. But whatever new forms of error are begotten of reaction, it is a comfort to believe that there are always some steps gained which are never lost. No man can look back on the history of modern civilisation without seeing that it presents the phenomena of development and growth. Nor can it be doubted, surely, that whatever may be the decline of particular Communities, the progress of mankind, on the whole, is a progress to higher and better things. And if this be true, no particular exceptions should shake our faith in the general rule that all safe progress depends on timely recognition being given to the natural developments of Thought. They can never be resisted in the end, and they are most liable to take erroneous directions when they are resisted long. For this is among the most certain of all the laws of Man's nature—that his conduct will in the main be guided by his moral and intellectual convictions. "All human society is grounded on a system of fundamental opinions." Such is the Law arrived at by the newest of modern Philosophies,[1] and it would be well if all its

[1] "The Positive Philosophy of Aug. Comte," by J. S. Mill, p. 101.

discoveries were as near the truth. This is the Law to which Christianity appeals, and in which its very roots are laid, when it asserts, as no other Religion has ever asserted, the power and virtue of Belief. And in this Law lies the error which those commit who imagine they can hold by the Ethics of Christianity, whilst regarding with comparative indifference its History and its Creed. This, too, is the Law which lends all their importance to the speculations of Philosophy. False conceptions of the truth, in apparently the most distant provinces of Thought, may and do relax the most powerful springs of action. Among these false conceptions of the truth, none are now more prevalent than those which concern the definition, and the function and the power of "Law." Instead of regarding the Constancy of Nature as incompatible with the energies of Will, we must learn to see in it the most powerful stimulus to inquiry, and the most cheering encouragement to exertion.

The superstition which saw in all natural phenomena the action of capricious Deities was not more irrational than the superstition which sees in them nothing but the action of Invariable Law. Men have been right and not wrong, when they saw in the facts of Nature the Variability of Adjustment even more clearly and more surely than they saw the Constancy of Force. They were right when they identified these phenomena with the phenomena of Mind. They were right when they regarded their own faculty of Contrivance as the nearest and truest analogy by which the Constitution of the Universe can be conceived and its order understood. They were right when they regarded its arrangements as susceptible of Change; and when they looked upon a change of Will as the efficient cause of other changes without number and without end. It was well to feel this by the force of Instinct; it is better still to be sure of it in the light of Reason. It is an immense satisfaction to know that the result of Logical Analysis does but confirm the testimony of Consciousness, and run parallel with the primeval Traditions of Belief. It is an unspeakable comfort that when we come to close quarters with this vision of Invariable Law seated on the Throne of Nature, we find it a phantom and a dream—a mere nightmare of ill-digested Thought, and of "God's great gift of speech abused." We are, after all, what we thought ourselves to be. Our freedom is a reality, and not a name. Our faculties have in truth the relations which they seem to have to the Economy of Nature. Their action is a real and substantial action on the Constitution and Course of things. The Laws of Nature were not appointed by the great Lawgiver to baffle His creatures in the sphere of Conduct, still less to confound them in the region of Belief. As parts of an Order of things too vast to be more than partly understood, they present, indeed, some difficulties which perplex the intellect, and a few also, it cannot be denied, which wring the heart. But, on the whole, they stand in harmonious relations with the Human Spirit. They come visibly from One pervading Mind, and express the authority of one enduring Kingdom. As regards the moral ends they serve, this, too, can be clearly seen, that the purpose of all Natural Laws is best fulfilled when they are made, as they can be made, the instruments of intelligent Will, and the servants of enlightened Conscience.

NOTES.

NOTE A.—PAGE 46.

THE article by Mr. Wallace in the *Journal of Science* (No. XVI.) for Oct. 1867, on "Creation by Law," implies much misconception of the whole drift and aim of the observations I have made on Mr. Darwin's "Theory of the Origin of Species."

Two separate questions arise in respect to that Theory:—

1st. Does it adequately explain the Physical Agencies by which new Forms have come to be?

2d. Does it, even if successful in this explanation, supplant or impair the argument for Purpose and Design, as founded both on the results and on the methods of Creation?

Of these questions, which are entirely distinct, the last is by far the most important of the two: and this second question is the one which I have chiefly dealt with in the "Reign of Law." As regards the first question, indeed, I have indicated my opinion that the explanation of Physical Agencies is very far from being complete, and that the hypothesis can always be driven back to some starting point where the very condition of things is assumed for which the theory professes to account. In this edition, and in answer to Mr. Wallace's challenge, I have added some farther indication of the difficulties which remain unsolved, and indeed untouched.

But my main argument has been addressed to the second question, and has been aimed chiefly at this conclusion—that even supposing the theory to be established, so far as it can go, it cannot affect or disturb the inseparable relation which exists between the intricate adjustments of Nature and Mental Purpose—as their sole conceivable origin and cause. In respect to this argument, Mr. Wallace virtually admits all I have maintained, when he says, "It is simply a question of how the Creator has worked."[1] I have said nothing of "incessant interference," of "continual rearrangement of details," or of "the *direct* action of the Mind and Will of the Creator." On the contrary, I have said that no purpose is ever attained in Nature

[1] *Journal of Science*, p. 473.

except by the enlistment of Laws as the means and instruments of attainment;[1] that although Law "is never present as a master, it appears never to have been absent as a servant,"[2] that we have "no certain reason to believe that God ever works otherwise than through the use of means;" or, in other words, through the instrumentality of those elementary forces or properties of matter and of mind which we call "Laws." The idea of "incessant interference" is one which I hold to be essentially erroneous. It involves the idea of natural forces being agencies altogether external to, and independent of, the Creative Mind. This is the very idea to which Mr. Wallace himself gives expression in its extremest form, when he limits the function of the Creator to that of so co-ordinating general laws "*at the first introduction* of life upon the earth,"[3] as that they shall work "of necessity" and "by themselves" the results we see. This is unquestionably the way in which they appear to us to work. It remains true that "no man hath seen God at any time." But even this conception does not make the word "contrivance" (to which Mr. Wallace objects) less applicable to the adjustments of Nature. On the contrary, it makes it more strictly and literally applicable than any other conception, because it likens the Creative process more closely to those methods

[1] P. 100. [2] P. 208. [3] *Journal of Science*, p. 474.

adopted by human ingenuity, to which the word contrivance specially refers. The highest efforts of that ingenuity are precisely those in which natural forces are made to work "necessarily" and "by themselves." Self-acting machines are the most ingenious machines of all. The self-action of the "governor" in a Steam-engine is the most beautiful of the contrivances by which the elementary expansive force of steam is made to do the work of Will. Mr. Wallace thinks it "an extraordinary idea to imagine the Creator of the Universe *contriving* the various complicated parts of an Orchis, as a mechanic might contrive an ingenious toy or puzzle."[1] But this is precisely the idea he himself supports, when he reduces the Creator's work to the first starting of the forces of organic life, and to the *foresight* merely of the consequences which must naturally and necessarily arise from their first co-ordination. This is an accurate description of the method in which a mechanic contrives the most ingenious self-operating machines. No doubt the idea of Omnipresence, which is the distinctive idea of God's work as distinguished from Man's work, is an idea which it is difficult for us to grasp or to keep steadily in view. I do not deny or dispute that "self-action" is and must be the aspect in which Nature presents herself to us. It could not be otherwise, unless the Invisible were to

[1] *Journal of Science*, p. 474.

become the object of sight and touch. But in proportion as we appreciate the infinite intricacy of Natural Adjustments, in the same proportion do we estimate the impossibility of conceiving them as the result of Mechanical Necessity, which indeed is an inadequate explanation even of our own methods of operation upon the material world.

Mr. Wallace's article is an excellent example of the arguments used in support of the Darwinian theory, both in their strength and in their weakness. Their strength lies in the hold they have of the idea and of the fact that Nature is one vast system of Invisible Forces in a condition of mutual adjustment. Their weakness lies in the idea that the methods of that adjustment can ever be explained as the result of mechanical necessity, or of the mere elementary properties of matter working "by themselves."

Note B.—Page 85.

Although the distinction I have made between Purpose as a general inference, and Purpose as a particular fact, is a distinction which seems to me to be clear enough when it is pointed out; yet it may be well to give some further illustrations here which could not be conveniently added to the text. What Positivists profess to insist upon is, that in describing a scientific fact, we shall not import into it ideas which it does not necessarily involve, and which are in the nature of inferences from the fact. What we, their opponents, have an equal right to insist on is this—that in describing scientific facts, the description must not *ex*clude any of the ideas which the facts do involve, and that the full and adequate description of those facts be not evaded in order to keep out an idea which the describer may choose to call an inference. Let us take an illustration. Let us suppose that we find a tube placed anyhow in such a position that we can look through it to the sky at night. We do so, and we see a star. The facts may be such that this description fully exhausts them. That the tube was intended to bear upon that particular star, or upon any star, would be a mere inference. But now let us suppose that, when we look again after some considerable interval of time, we find that the tube still bears upon the same star, and let us further suppose the same experiment repeated with the same result during some hours, then we should not describe the fact fully by simply stating that the tube bore upon the star. It would be necessary, in order to exhaust the facts, to say that the tube was *so adjusted as to follow* the apparent motion of the sidereal heavens, and so to counteract the natural effects of the earth's motion as to keep its axis always upon the same star. Here instantly we have the language of intention, because the idea of intention is inseparable from the facts. We might know nothing of the method by which this adjustment is achieved—nothing more of the Mind that had devised the method than the bare fact of the intention. But that bare fact is an essential part of the observed phenomena. And the same argument applies to the mechanism—if that also were discovered—by which the adjustment is effected between the axis of the tube and the apparent motion of the star. That mechanism could not be fully described unless it were described as a mechanism *so contrived as to bring about* the adjustment which is actually effected.

Take again another case, from the organic world. A calf, or any other young animal, discovers by smell or by accident, the fact that milk is contained inside a skin or bag, and that, by applying its mouth or its tongue to some opening, it can get at the milk. The whole fact in this case is exhausted when we say that the calf gets the milk. It is no part of this fact that the calf was intended to get it. But when a calf goes for milk to its mother's udder—when the lacteal glands of the cow are recognised as an apparatus for secreting that milk, and the teat for delivering it,—then the facts are not exhausted, the scientific description is not complete or truthful, unless we use language importing this adjustment of apparatus to Purpose in the plan by which nourishment is afforded to the young in all mammalia. This idea cannot be expelled from science as a mere "inference," except on the same arguments of bad metaphysics, as I hold them to be, by which also the existence of Matter and of an External World are referred to the same category.

NOTE C.—PAGE 86.

In illustration of the assertion in the text, that the relations of Number, which are the very basis of all "verifiable" knowledge, may be reduced by similar arguments to mere creations of the Mind, I may here remind the reader of the passage which relates to this subject in the famous argument of Berkeley :—

"That number is entirely the creature of the mind, even though the other qualities be allowed to exist without, will be evident to whoever considers that the same thing bears a different denomination of number as the mind views it with different respects. Thus the same extension is one, or three, or thirty-six, according as the mind considers it with reference to a yard, a foot, or an inch. Number is so visibly relative and dependent on men's understanding, that it is strange to think how any one should give it an absolute existence without the mind. We say one book, one page, one line; all these are equally units, though some contain several of the others. And in each instance it is plain the unit relates to some particular combination of ideas arbitrarily put together by the mind."—*Prin. of Hum. Knowl.*, Part I. § xii.

Note D.—Page 204.

Mr. Mill, in his "System of Logic" (Book I. c. iii., §§ 6, 7, 8), has told us that both of Bodies and of Minds, "philosophers have at length provided us with a definition which seems unexceptionable." As regards Body, this definition is—"The external cause, and (according to the more reasonable opinion) the unknown external cause, to which we refer our sensations." This definition, though very defective, is at least not erroneous. It is necessary, however, to observe that the word "unknown" cannot be accurately predicated of that respecting which the very terms of the definition imply that something is known. The definition of Body is the definition of that which is known respecting it. Three things are involved in this definition, as known respecting Body;—these are (1) Externality, (2) Extension, and (3) Causation—that is to say, the power of causing or exciting sensations in sentient beings. Or perhaps these three items of knowledge may be merged in one—the knowledge of Force acting from outside upon us, and exciting sensations in us. But this is knowledge. When the word "unknown" therefore is inserted in the terms of the definition given by Mr. Mill, it can only mean that other things still remain to be known respecting the nature and properties of Body. In this sense—that is, when translated into "partially known"—no philosopher would deny the correctness of its application. The definition of both Body and Mind is given by Mr. Mill in another passage, which also, so far as it goes, is unexceptionable. "As Body is understood to be the mysterious something which excites the Mind to feel, so Mind is the mysterious something which feels and thinks." The same two fundamental ideas of Externality and Causation are here also implicitly and inextricably involved. Mr. Mill adds that the farther discussion of this question belongs not to Logic but to Metaphysics, to which science he leaves it.

Two chapters, accordingly, in Mr. Mill's "Examination of Hamilton" (xi. and xii.), are devoted to a discussion of the "Psychological Theory of the Belief in an External World," and of the question how far the same theory may or may not be also applicable to Mind. The conclusion to which I have referred in the text is the conclusion defended in the first of these chapters. It is the conclusion of a Pure Idealism—an Idealism much more extreme than the theory of Berkeley. It is true that Berkeley denied the existence of Matter as a thing apart from Mind—not, be it observed, as a thing apart from *our minds*, but as a thing apart from *some mind.* But this was only because he sublimed it into the action of another Spirit upon our own. In his system the idea of Causation was tenaciously retained. The very essence of his argument was that our ideas must have a cause—"some cause whereon they depend, and which produces them and changes them." As this cause could not be the ideas themselves (which ideas are all that we know of matter), "it remained that the cause of Ideas is an incorporeal active substance or spirit."[1] This argument is repeated in several forms, as again where he says,[2] that men were "conscious that they were not the authors of their own sensations, which they evidently knew were imprinted *from without*, and which therefore must have some cause distinct from the Minds on which they were imprinted." But the Psychological Theory of Mr. Mill involves all the weak points of the Berkeleian theory with none of its strength. Mr. Mill's formula is expressly framed so as to eliminate as much as possible the idea of Causation, and to keep out of sight the connexion which exists between our sensations and that which excites them. The attempt, indeed, is not successful, because Mr. Mill cannot express himself

[1] "Prin. of Hum. Know." § xxvi. [2] Ibid. § lvi.

through many consecutive sentences without assuming the very ideas which he is trying to account for as a mere product of more elementary conceptions. This has been shown clearly, and with abundant illustration, in Dr. M'Cosh's "Examination of Mr. J. S. Mill's Philosophy." Mr. Mill pleads upon this point that he must use common language, but that the whole of this language has its own special meaning under the Psychological Theory as well as under the common Realistic Theory. This may be true; but there are certain words which must have the same meaning under all theories; and, in spite of his efforts, he is compelled to employ words which show that neither he nor any one else can maintain consistently a purely subjective conception of Matter,—that is to say, a conception which dispenses with an external agency or force. He says, that "almost all philosophers, who have narrowly examined the subject, have decided that Substance need only be postulated as a *support* for phenomena, or as a bond of connexion to hold a group or series of *otherwise* unconnected phenomena together." Mr. Mill goes on with much simplicity: "Let us only then *think away* the support, and suppose the phenomena to remain, and to be held together in the same groups and series by *some other agency*, or without any agency *but an internal law*—and every consequence follows without Substance, for the sake of which Substance is assumed."[1] The demand here made upon us, to "*think away*" the support of phenomena, is certainly made less formidable when, in the next breath, we are told to *think it back again* under another form of words, as "another agency," or as an "internal law."

The same vain attempt to get behind ultimate ideas may be traced in the word "Permanent," with which Mr. Mill qualifies Matter considered as "A Possibility of Sensation." The new formula is "A *Permanent* Possibility of Sensation." Why permanent? Permanent means enduring. But what has the element of Time to do with it? The percipient minds are not permanent, so far as the sensations of their existing organism is concerned. In what sense, then, are the "Possibilities of Sensation" permanent? What is it that is described as permanent? Not the sensations—not the individual sentient beings. What then? Clearly the Power or Agency which causes, or is capable of exciting sensations in organisms that are, or that are to be. Here, then, we have the ideas of Externality and of Causation brought back under the covering of Time. "What is it we mean," asks Mr. Mill, "when we say that the object we perceive is external to us, and not a part of our own

[1] Appendix to chaps. xi. and xii. "Mill on Hamilton," 6th ed. p. 246.

thoughts?" The reply to this question, in the first Edition, ran as follows: "We mean that there *is in our perceptions* something which exists when we are not thinking of it, which existed before we ever thought of it, *and would exist if we were annihilated.*" In the recent Edition, this reply has been altered so as to avoid the obvious absurdity of supposing that things which are conceived to exist only "in our perceptions," could nevertheless continue to exist "if we were annihilated." Accordingly the reply now runs thus: "We mean that there is *concerned in* our perceptions," &c. Yes; but how concerned? As an exciting Force or producing Cause, and in no other way. Similar observations apply to the word "Possibility," as applied in Mr. Mill's Psychological Theory. Possible can only mean Potential. A Possibility of sensation must mean a Potential Cause of sensation. And here, again, we have the same fundamental ideas involved in the very language by which it is attempted to evade them.

Mr. Mill appears to me to be equally unsuccessful in starting fairly in this Psychological Theory—that is to say, in the definition of postulates which steer clear of involving the very ideas for which he professes to account. His first postulate is that the Human Mind is capable of Expectation. Certainly; but what does Expectation involve? It involves acts of Memory, and of Comparison, and of Reason. In particular, it involves, or at least he is not entitled to deny that it may involve, the intuitive belief that actual sensations already experienced arose from an external cause, and that the same cause is capable of exciting them again. My belief is that the mind cannot place itself in the attitude of expectation without the presence of ideas which involve the whole question in dispute.

I am disposed, therefore, to agree with Mr. Mill, that the existence of an external material world cannot be proved;—just in the same sense, and for the same reason, that the proposition—"Things that are equal to the same thing are equal to one another," cannot be proved.

Mr. Mill thinks that, though the existence of an external Material world cannot be proved, an external Immaterial world can be proved—that is to say, the existence of other minds can be proved. I think he only succeeds in showing that our belief in this existence can be confirmed by corroborative evidence. But such corroboration and confirmation is equally available in support of the belief in the existence of Matter, considered as an External Cause of sensation. The truth is, our knowledge of other minds is only reached through our previous knowledge of Matter, and of the impressions it makes upon us. My own mind, as well as the mind of all the beings around me, is, or seems to be, inseparably connected with a Material Organization, and there are no manifestations of Mind which do not come to me directly or indirectly through material signs.

Mr. Mill has often warned us, and I accept the warning, against the system of discussing metaphysical questions, under the threat, as it were, that the conclusions to which we are opposed are inconsistent with some one or more Theological Beliefs. We know that the Ideal Theory, in the form at least which it took in the hands of Berkeley, was put forward in the interests of Religion. "The existence of matter, or bodies unperceived, has not only been the main support of Atheists and Fatalists, but on the same principle doth Idolatry likewise, in all its various forms, depend. Did men but consider that the sun, moon, and stars, and every other object of the senses, are only so many sensations in their own minds, which have no other existence but barely being perceived, doubtless they would never fall down and worship their own *Ideas*, but rather address homage to that Eternal Invisible Mind which produces and sustains all things."[1] Such was the animating principle of the Bishop of Cloyne's famous speculation. I confess I have a profound distrust of

1 "Prin. of Hum. Knowl." § xciv.

all attempts to found the teachings of Faith upon the principles of Scepticism. I am not tempted, in order to escape the danger of Materialism, to deny the existence of that, which I know by my own structure, and by the structure of all around me, to be different from Mind. I am content to understand the world as my own faculties have been co-ordinated with external things to reveal those things to me. I look in my friend's face, and I see the expression of power, and of intellect, and of goodness. These are attributes of Mind. I do not know *how* these attributes can be shown forth so evidently in the colours and in the lines of flesh and blood. But I do not try to persuade myself that his hand or his face are the same things, either with the perceptions which they excite in me, or with the emotions which they express in him. I do not pretend to understand the nature of the connexion between these Material Forms and the qualities of Mind. But after their own diverse kinds and measures they are both equally "real" to me. I will not deceive myself by verbal quibbles—pretending to be able to stand outside myself, and to prove by reason that the very tools with which reason works are rotten in her hands. There is but one sentence in these two chapters of Mr. Mill's work which conveys any really important truth. In regard to the existence of Matter, as well as in regard to the nature of Memory and of Mind, we may indeed well say with him: "By far the wisest thing we can do is to accept the inexplicable facts, without any theory how it takes place; and when we are obliged to speak of them in terms which assume a theory, to use them with a reservation as to their meaning."

NOTE E.—PAGE 299.

When I wrote this passage in the text, I had not read a curious note by Sir W. Hamilton in his edition of Reid's Works.[1] It is almost droll in its confession of the puzzling significance of such facts, in respect to animals, as those I have referred to. "Nothing in the compass of inductive reasoning appears more satisfactory than Berkeley's Demonstration of the necessity and manner of our learning, by a slow process of observation in comparison alone, the connexion between the perception of vision and touch, and, in general, all that relates to the distance and real magnitude of external things. But although the same necessity seems in theory equally incumbent on the lower animals as on man, yet this theory is provokingly (!)—and that by the most manifest experience—found totally at fault with regard to them; for we find that all the animals who possess at birth the power of regulated motion (and these are those only through whom the truth of the theory can be brought to the test of a decisive experiment), possess also from birth the whole apprehension of distance, &c. which they are

[1] Vol. I. p. 182, note.

ever known to exhibit. The solution of this difference by a resort to instinct is unsatisfactory; for instinct is in fact an occult principle—a kind of natural revelation,—and the hypothesis of instinct, therefore, only a confession of our ignorance: and at the same time, if instinct be allowed in the lower animals, how can we determine whether and how far instinct may not in like manner operate to the same result in man?"

Well might Sir W. Hamilton ask this question. It is one which Philosophers will find it hard to answer. My own conviction is that more than half the "inductive reasoning" by which men have pretended to account for their intuitive perceptions is altogether unsound. Man, besides being man, is also an animal—and through his animal organisation the mechanics of his mind are to a large extent regulated on the same principles which regulate the lower Intelligences around him—that is to say, by processes unconsciously pursued. This is the proper definition of operations which are Instinctive; and, as Sir W. Hamilton observes, they may best be conceived as the result of "Natural revelation."

Note F.—Page 304.

In the number of the *Dublin Review* for April 1867, there is an article on "Science, Prayer, Free Will, and Miracles," in which some portions of this work are criticised. With much of that criticism I have every reason to be satisfied. The main objection taken may be stated in two sentences. I have said that (under certain limitations as to the meaning of the words) the "abstract predictability of human conduct" may be admitted without involving the denial of anything essential to the doctrine of Free Will. Dr. Ward denounces this concession as absolutely fatal to that doctrine, and maintains that in making such a concession, as well as in other more direct forms of statement, my view comes to be "precisely identical with Mr. Mill's," which, nevertheless, I am "professing" to oppose. This position he proceeds to support by an elaborate argument, which I shall here examine with all the care due to the gravity of the question raised, and to the duty of using no language upon such a subject which is not justified by as much precision of thought as is attainable in regard to it.

As Dr Ward speaks upon this subject with some

warmth of feeling, perhaps I may explain at once that he is mistaken in supposing that I am "a Calvinist," in the sense of holding "the Necessitarian Doctrine." I hold the doctrine of Free Will in the only sense in which it is to me intelligible. I set the highest value upon it; and in the result, though not in this particular argument, I believe I agree with Dr. Ward himself. I am willing to accept without reservation the definition which he quotes from certain Jesuit theologians: "Potentia libera est quæ, positis omnibus requisitis ad agendum, potest agere et non agere." But Dr. Ward does not seem to observe that in this definition the whole question in dispute may be covered under its contingent clause. Everything depends on the further definition to be given of "all the requisites for action." Is, or is not, the condition of the mind itself to be considered as one of those "requisites?" Is knowledge, and the possession of those motives which knowledge gives, a "requisite" or not? Do the "requisites" intended, by the Jesuit definition, refer to nothing more than the presence or absence of physical constraint? The truth is that such abstract definitions go very little way in explanation of themselves. I have asserted the freedom of the Will under several forms of statement which are much more explicit, because much more full and more detailed. Thus, I have said, "Among the motives which act upon mind, Man has a selecting power. He can, as it were, stand out from among them—look down from above them—compare them among each other, and bring them to the test of conscience." This is freedom, if there be such a thing conceivable in thought. But Dr. Ward's impetuous zeal in favour of Free Will blinds him to certain truths which are perfectly compatible with this doctrine, and which not only must be admitted, but must be claimed, if we are ever to wield against Necessitarians the weapon of analysis. The principal object of this work has been to show that, in so far as science has successfully established in physics the idea of the Reign of Law, that idea does not affect or traverse the Reign of Mind, and the supremacy of Purpose. In like manner, I think it can be shown that in so far as Psychology can successfully establish the idea of Causation, as applicable to Mind, that idea is perfectly compatible with the true freedom of the Will. Dr. Ward says very truly, that the Necessitarian doctrine has in all ages been embraced by many powerful minds. This indicates that, in so far as it is false at all, its falseness probably depends on some partial aspects of truth mingled with the fallacies of definition. My own opinion unquestionably is, that when Necessitarians have been compelled to disown and abjure the idea of compulsion, their doctrine ceases to be the doctrine of Necessity at all in any legitimate sense of the word. What I mean by freedom is freedom from compulsion, and nothing else. When I say that the Will is free, I do not mean that its movements can be separated from the inducements internal and external under which it moves. But then I insist that "Motive" shall have the widest meaning—that it shall include such motives, evolved out of the very constitution of the mind itself, as "Love, and Reverence, and Gratitude, and Hunger after Knowledge and Desire of Truth." Of course this is not given as a complete list, but only a sample of the things which must be claimed as "motives." In this sense, not only is the determining power of Motive inseparable from the very idea of Mind; but the higher is the quality of a mind, the more certain and definite will be the motives of its action. By some strange confusion of thought, Dr. Ward seems to regard with horror the idea of the Will being regarded as part of "the constitution of the mind." This is a mere question of words. But if by Will, Dr. Ward does not understand a particular power of mind, I do not know what he means. To analyse Mind at all, we must of course consider its different powers as separate from each other; but it does not the less remain true that they are all parts of one whole. In this point of view, Dr. Ward's own definition of Free Will is far from clear. "The Will, we maintain, has a certain power of deciding *for itself* what weight it shall attach to motives." Certainly, if the Will be understood as including the Deliberative Faculty whose function it is to "weigh." But in coming to this decision it must be guided by something, which something may always itself be resolved into another "motive."

And if this appears to be a mere play on words, I grant it. It is the very point and object of my argument to show that in so far as the Necessitarian doctrine has any apparent force, it does depend on mere ambiguities of language. For example—exclude from the word "motive" all the influences which come upon the spirit from the mind itself—from conscience, from the action upon it of another Spirit, human or divine—confine the word "motive" (as many do, tacitly by implication, though not consciously) to that class of motives which come from external and material things,—in short, confine it to the appetites or desires, then it absolutely ceases to be true that the Will is determined by "motives." On the other hand, include in the word "motive" *all* that can ever influence the mind, whether from within or from without, then it ceases as absolutely to be true that the Will can ever be "free" from such motives. But then, in this sense, the Necessitarian doctrine resolves itself, as Mr. Mansel says, into the identical proposition that "the prevailing motive prevails." It becomes perfectly harmless, because in reality perfectly unmeaning. Dr. Ward is very indignant that I should represent my view as "a mere truism." But it is not my own view, but the Necessitarian doctrine, when thus reduced by analysis to its real value, which I have represented as a mere truism.

All these fallacies and confusions of thought arise, in my opinion, from neglect of the fact that freedom has no absolute, but only a relative meaning. Freedom can only mean "the not being bound," and bonds can only consist in something binding. Freedom of the Will can only mean that the Will is free from compulsion. If Necessity does not mean compulsion, it either means nothing at all, or nothing inconsistent with freedom when properly defined and understood.

We now come to what is called the "Abstract predictability of human conduct." This is the phrase into which Mr. Mill retreats, as containing the residuum of truth which still belongs to the Necessitarian doctrine after it has abjured the idea of compulsion. It is not my phrase, or one which I approve of, because it involves a great number of assumptions which lie, as it were, concealed within it. But I adhere to the opinion that, when strictly defined, the idea it involves is perfectly capable of being reconciled with the freedom of the Will. The truth is, that it is capable of being resolved into the same identical proposition as the Necessitarian doctrine in other forms.

If by "abstract predictability" is meant that prediction would be possible under the conditions of complete, universal, and perfect knowledge, I do not see either how it can be denied, or to what purpose it can be affirmed. The proposition is that, if ALL the conditions were known which determine the Will in deciding for itself, or "in giving weight to motives," the result of that decision would thereby become also known. Of the Necessitarian doctrine expressed in this general form, I have said, and I repeat, that it is "very like a truism." But if it is useless as an affirmation, it is at least not capable of denial. Dr. Ward, however, does deny it, and supports his denial by reasoning which is clearly untenable, as an admission made by himself will show. His idea seems to be, that no "predictable" conduct can be "free;" that nothing which can be abstractedly foreseen can be the result of freedom. But Dr. Ward does not, and cannot maintain this view consistently. He admits that, "taking any given man at any given moment, there are certain things so good, and certain things so bad, that we may infallibly calculate he will do neither the one nor the other." Would

Dr. Ward then admit that as regards those "very bad," and those "very good," deeds, this man is not "free?" Or does he think he escapes this difficulty, by putting the man's conduct in the negative instead of the positive form? As regards the action of the Will, no such distinction is of any avail. The not doing one thing is the doing of another. The not doing a very good deed, which he has power to do ("positis omnibus requisitis ad agendum"), is willing not to do it. The not doing a very bad deed is willing to do something else. If then the conduct of the man in those cases "can be calculated with perfect certainty," it is so calculable only because knowledge of his character is convertible into knowledge of the manner in which his Will is sure to act. Is it not then a clear violation, both of the ordinary and of the philosophical use of language, to say that a man is not "free" to do a very bad act, because we know certainly beforehand that his character, and the motives on which he habitually acts, will prevent him from doing it?

But then Dr. Ward proceeds to argue that though infallible calculation may be possible in respect to deeds very good, and very bad, it will not be possible in regard to deeds only a little good and a little bad. But how does this greater difficulty arise? Is it not because the number of motives telling on the Will is greater, more nicely balanced, and therefore less known? And is not this difference precisely the kind of difference which would disappear, if he could pass from knowledge which is partial only, to knowledge which is complete and absolute? But whatever difficulty may arise from imperfect knowledge is (as I understand the phrase) eliminated by the word "abstract," as qualifying "predictability." No one asserts that prediction can be founded on partial knowledge. But the question raised is whether even perfect knowledge of all the elements of motive and of character can render predictable the conduct of a really Free Agent. The question is one involving a logical principle, which, if applicable to the conduct of a Free Agent in any case, must be equally applicable to his conduct in all cases. If it is abuse of terms, or a confusion of thought, to affirm that a man's Will is not free to do or not to do very bad actions, because we can calculate infallibly the decision of his Will in regard to them, it must be equally fallacious to affirm that his Will would not be free in regard to lesser degrees of vice and virtue if, in like manner, we were able from perfect knowledge of his character to predict his conduct also in respect to them. The doctrine of Free Will, like every other doctrine of Mental Science, can only be defended by clear definitions of what it is. Its defenders have in my opinion established their case when they have compelled Necessitarians to discard the idea of compulsion. All attempts to deny that the Will is determined by "motives" are futile, and only result in giving a seeming victory to those who have in reality been defeated.

In order to illustrate what I mean, I will suppose a particular case; and to comply with the conditions of Dr. Ward's argument, it shall be a case where no determination, either very good or very bad, is involved. I will suppose that in arguing with a friend on the subject of Free Will, a plate of oranges is offered to me. My friend tells me that "he knows which of these oranges I shall choose." I tell him he cannot possibly know this—that my Will is free, and therefore he cannot predict my choice. He insists upon it that he can. I then observe that one orange has a smoother skin than the others, and is of a deeper yellow colour. I then recollect that I had once mentioned in my friend's hearing that I considered a pale colour, or a rough skin, as indications of sour or tasteless oranges; and remembering this fact, I at once perceive that my friend is calculating my conduct from a motive which, as he knows, does habitually determine my choice of oranges. I am conscious also that in this particular case I should have been so determined—if this dispute had not arisen. But in order to show my friend that my Will is really free from this power of "motive," I determine to exert that freedom by choosing the palest, or the roughest orange in the plate, and I accordingly do so. This is an assertion of my Free Will—a practical denial of the doctrine that I am the slave of "motives." But is it not clear in this case, that my conduct has been determined after all only by another and a stronger motive than the one which usually acts with me in the matter—even by the motive of proving to my friend that he was wrong, and that I was right?—a motive which is strong with all men, and is supposed to have special attractions for a Scotchman. And is it not equally clear, that if my friend had had more perfect knowledge of my character, and had known that I recollected the former conversation, and could therefore guess the grounds of his prediction, he might, and would have been able to foresee correctly the new motive which had thus arisen to overpower the other? And finally, is it not equally evident that, if he had been able by this extraordinary sagacity to predict my choice correctly, the correctness of that prediction would not have implied the existence of any constraint on the freedom of my Will, but, on the contrary, would have been founded on his knowledge of my freedom to pass from the old motive and to give effect to the new one?

A thousand different examples of the same kind might

be given. That on which the Will finally determines to act may always be called, and is always properly called, a motive. And this is observable in respect to the whole question, that precisely in proportion to the high qualities of any given mind—in proportion to its intellectual power and its moral strength—in proportion to its keen insight into the causes and tendencies of things, and its appreciation of truth and righteousness—in the same proportion will the distinction vanish in its eyes between things "very bad," and things only a little bad. In the same proportion, therefore, will its own conduct be guided by definite and certain motives: in the same proportion, finally, will that conduct become predictable, because in the exercise of its freedom it is governed by moral laws which never change.

Note G.—Page 306.

Mr. Mahaffy, in his article in the *Contemporary Review*, has taken objection to the breadth of meaning which I have given in this passage to the word "motive." He says, I have "surely fused together two opposite theories under the ambiguous meaning of *motive*." This is precisely what I have done, and what I meant to do. I adopt all that I consider to be true in the so-called Necessitarian Doctrine, which, when cleared from the idea of compulsion, is no doctrine of necessity at all. The residuum of truth is, that the Will must always act on some motive. I have taken also all that is of any value in the Doctrine of Free Will, which is—that among the "motives" of the mind must be reckoned those inducements which arise out of its higher, as well as out of its lower faculties. But Mr. Mahaffy is, in my opinion, clearly wrong when he objects to the word "motive" being employed with this breadth of meaning. His objection indeed is explained to be that the word "motive" ought not to be applied to the "action of the Will upon itself." But I have not so applied it, because I have no notion what "the action of the Will upon itself" means. Mr. Mahaffy gives a farther explanation of this expression, when he speaks of the Will "creating principles of action for itself." But I deny altogether that the "creating" of anything is the function of the Will. It is by an act of Will that we fix our attention upon any given motive, or turn, on the contrary, our attention from it. But if we are to analyse the mind at all, if for the convenience of thought and of discussion, we are to divide its inseparable Unity into different powers, we must make the division as logical as we can—that is, as consistent as possible with definite ideas of distinct mental functions. In considering the Will as a separate Power, we must strictly confine it to what may be called the Executive of the Mind. In this light it would be altogether incorrect to ascribe the "creation" of any motive to the Will. Motives of all kinds, both the highest and the lowest, may rise, and do rise unbidden in the mind. It is by an act of Will that we summon different motives to the presence of the Deliberative Faculties, that we cherish one and dismiss another, or determine to spend thought and time in making our choice between motives which are conflicting. But the Will cannot with accuracy be said to be the creator of motives. Intellectual and moral conceptions, held together by the bonds of Memory, are the fountains from which the highest motives come. Mr. Mahaffy admits that "anything brought to bear upon the Will *from without itself*, even from the intellectual part of the mind, is a motive." But according to my definition of the Will all motives come equally from outside the Will, and assuredly I see no ground for the distinction Mr. Mahaffy seems to draw between the Intellectual and the Moral faculties. In denying the name of motive to those inducements which come from the affections or from the sense of right and wrong, he imposes a restriction on the meaning of the word which is not less inconsistent with common usage than with philosophical accuracy. Affection and gratitude, the love of man and the love of God, are all surely "motives" in the most proper sense of the term. Yet Mr. Mahaffy asks, "Is it not an abuse of language to say that the man who resists temptation by creating within his breast a strong feeling of moral responsibility is determined by motives?" To this question I reply at once (passing over the question of the "creation" of motives) that it is no abuse of language, but, on the contrary, the employment of language in its natural and ordinary sense. On what principle is the love of knowledge (being intellectual) to be called a motive, if the love of God is not? On what principle is a desire of producing physical results to be called a motive, if the desire of attaining moral ends is denied the name? No such distinction is tenable in a philosophical point of view, and no such distinction is known in the usual and familiar employment of the word "motive." Mr. Mahaffy, however, in making this objection, has put his finger upon the point on which the whole discussion turns. Like many other metaphysical questions, it depends almost entirely on a definition of terms. If the word "motive" be arbitrarily limited to mental affections of one or two particular kinds, if it be confined to the lower appetites and desires, or even if it be extended to the higher appetites of the Intellect, whilst it is denied to the inducements of morality, of conscience, of Religion,—then it ceases to be true that the mind is determined by motives alone. The result is that the so-called "Necessitarian" Doctrine, in so far as it is true at all, must not only exclude the idea of compulsion, but it must include all that class of inducements, on the pre-existence of which, and on the power of choice among them, the responsibility of the Will depends.

The view presented in the text of the great question of Necessity and Free Will does fuse together some portions of the two opposite Theories which have so long divided men's minds regarding it. But in this fusion I do but follow the process pursued by Dante in a profound and beautiful passage of his "Purgatorio" (Canto 18th). To Necessity he ascribes the existence and the power of Motive. Motives arise out of the relations pre-established between the Human Spirit and all the Influences by which it is surrounded. No other account can be given of them. Dante sees no difficulty, as some modern defenders of the Free Will doctrine do, in comparing the ultimate nature and origin of all our mental desires with the nature and origin, equally inexplicable, of the Instincts of the lower animals. Hear the lines, not less musical in sense than they are in sound—

> "Però, là onde venga lo intelletto
> Delle prime notizie, uomo non sape
> E de' primi appetibili l'affetto;
> Che sono in voi, sì come studio in ape
> Di far lo mele."

To Free Will he ascribes the power of Counsel—of deliberation and of choice among the motives which thus arise from the very nature and constitution of the Mind. This power guards the "Threshold of Assent:"—

> "Innata v'è la virtù che consiglia,
> E dell' assenso de' tener la soglia."

On this power depends the responsibility of conduct:—

> "Quest' è'l principio là onde si piglia
> Cagion di meritare in voi, secondo
> Che buoni amori o rei accoglie e viglia.

The passage closes with these beautiful and striking lines :—

> "Color che ragionando andaro al fondo
> S'accorser d'esta innata libertate;
> Però moralità lasciaro al mondo.
> Onde pognam che di necessitate
> Surga ogni amor, che dentro a voi s'accende;
> Di ritenerlo è in voi la potestate."

It would be difficult to give so much philosophy in fewer words.

Note H.—Page 309.

In the last edition of Mr. Mill's work (1867), he has made an addition to the sentence quoted in the text, so that it now runs thus :—" I deny it as strenuously as any one, *in the case of human volitions, but I deny it just as much of all other phenomena.*" If Mr. Mill means, by this addition, to imply that he can deny compulsion (for example) in the behaviour of a billiard-ball, when it is struck, "just as much" as he can deny it, of the behaviour of a man when he is insulted, he renders his previous explanation valueless, and restores again to the doctrine of Necessity that very element of meaning which he professes to disclaim. Compulsion is predicable of the effects of Physical Force exerted upon Matter, in a sense in which it is not predicable of the effects of Moral or Intellectual inducements exerted upon Mind. This is precisely the distinction which Necessitarians are perpetually confounding ; and so long as they do confound it, their doctrine is justly open to the objection implied in the name usually assigned to it. Even if it be true, as Mr. Mill holds, that we have no other idea of Physical Causation than that of uniform and invariable sequence,—no idea of Necessity in Causation,—still it remains true that Compulsion is, apparently to us, involved in the effects of Physical Forces upon Matter, in a sense in which it is not involved in the effects of "Motive" upon Mind.

INDEX.

B.

D.

E.

F.

G.

H.

I.

J.

K.

L.

M.

P.

T.

U.

V.

W

Z.

www.ingramcontent.com/pod-product-compliance
Lightning Source LLC
LaVergne TN
LVHW020914110826
845150LV00004B/671

9781425556273